The Blurring of Boundaries in Bioscientific Discourse

Jana Tomašovičová (ed.)

Logos Verlag Berlin

λογος

Bibliografische Information der Deutschen Nationalbibliothek

Die Deutsche Nationalbibliothek verzeichnet diese Publikation in der Deutschen Nationalbibliografie; detaillierte bibliografische Daten sind im Internet über http://dnb.d-nb.de abrufbar.

The Blurring of Boundaries in Bioscientific Discourse
Jana Tomašovičová (ed.)

This publication has been peer reviewed.
Reviewers:
Prof. Mgr. et Mgr. Andrej Démuth, PhD.
Prof. PhDr. Matúš Porubjak, PhD.
Translation and proofreading:
Mgr. John Peter Butler Barrer, PhD.

Cover design and layout:
Paula Klarck

This work was supported by VEGA project 1/0563/18 and the Faculty of Arts of the University of Ss. Cyril and Methodius in Trnava.

ISBN 978-3-8325-5422-4

Logos Verlag Berlin GmbH
Georg-Knorr-Str. 4, Geb. 10,
D-12681 Berlin

Tel.: +49 (0)30 / 42 85 10 90
Fax: +49 (0)30 / 42 85 10 92

http://www.logos-verlag.de

Contents

Introduction

The gradual blurring of previously valid boundaries can be observed in almost all areas of social, cultural, and political life. Drawing boundaries allows us to establish a certain order by which we direct our actions; however, any order is constantly confronted with the current state of knowledge, leadership, and the needs and goals of society. In this confrontation, established boundaries, frameworks of interpretation, and standards of judgment are either confirmed or placed in doubt. Currently, boundaries are shifting under the influence of dynamic developments in bioscientific knowledge. New technologies have revealed previously unknown and invisible parts of the human body and made it visible at the molecular level, revealing in turn more detailed structures and arrangements than those which were previously available. As a result, current knowledge is being refined, expanded, and even fundamentally questioned in many ways. Following Michel Foucault, we can say that modern technologies allow us to discover new specific spaces – "heterotopias"[1] – which are quite different and distinct from usual spaces and which therefore disturb us. They challenge the established order of things, our sets of relationships, and our preferred forms of thinking. Contemporary bioscientific knowledge presents these new dimensions, which are uncovered thanks to new technologies; as result, established ways of interpreting and reinterpreting the world are gradually being disrupted and obscured. For philosophers, this is a most timely challenge that calls for a deeper analysis.

The aim of this book is to explore the shifts and blurring of boundaries in several areas with a specific focus on current bioscientific discourse. The authors of this book's chapters trace the shifting of boundaries in terms of the gradual blurring of the validity of established concepts, interpretive schemes, and standards of judgment, which are analysed from ontological, gnoseological, ethical, and social perspectives. At the same time, they also map the blurring of boundaries in terms of the interdisciplinary crossing of boundaries between various scientific and artistic disciplines.

In the first chapter, Juraj Odorčák discusses the idea of the complete technological automation of human reproduction, which he calls "robogenesis". In addition to focusing on conceptual and technological assumptions, he primarily explores its ontological aspects from the position of humanism, transhumanism, and posthumanism. According to Odorčák, the ontological ambiguity of

[1] Foucault, Michel. 1986. "Of Other Spaces." Trans. by Jay Miskowiec. *Diacritics* 16, 1: 22–27. DOI: https://doi.org/10.2307/464648.

robogenesis implies a certain inadequacy of the usual dichotomy between natural and artificial human reproduction; there is also a third factor – namely, culture – that plays a fundamental role in robogenesis and which organizes itself without human intervention (AI, automation). The specificity of robogenesis thus lies in the distinction between nature, culture, and "natureculture", which presents questions about the ontological status of the technologically created human being.

In the second chapter, Andrej Rozemberg explores the ontological problems of determining a person's identity. He uses John Locke's memory theory of personal identity to illustrate the theoretical difficulties that can arise from confusing identity with continuity. According to Rozemberg, the main problem with memory theory is not its circularity but rather the fact that it does not sufficiently reflect the phenomenological level of remembering and forgetting that allows for a distinction between memories and the remembering subject. He justifies why he considers the diachronic self – which he interprets as the "substantive self" due to several phenomenological specificities – as a more appropriate starting point in solving the problem of transtemporal personal identity.

The gnoseological problem of identifying the origin of rules is examined in the third chapter, which is written by Tomáš Čana. Based on an analysis of Ludwig Wittgenstein's later work, he asks about the basis of our activity in following (or not following) rules. Can the origins of the rules and criteria upon which we make decisions in life be rationally understood and explained? Čana agrees with Wittgenstein's genealogy of normativity, but he disputes its corollary, ultimately presenting an anti-theoretical position. Following Michael Dummett and Crispin Wright, he argues that recognizing the limits of the theoretical is not a reason to abandon the project of developing a theory of the origin of criteria.

When examining ethical issues in biomedicine, there is often a focus on respecting the principle of autonomy and the autonomous actions of the individual. The theory of autonomous agency proposed by Tom L. Beauchamp and James F. Childress is now considered the standard conception of autonomy in bioethics. Martin Zielina and a broader team of authors have investigated the extent to which this theoretical concept is realistically translated into practice, applying it to the doctor–patient relationship. The authors present the results of their unique empirical research, which was conducted in Czech hospitals. They observed the fulfilment of the concept of default autonomy in ethically controversial cases. They found that only 21 percent of the cases fulfilled all the criteria of default autonomy. A detailed analysis of this research is presented in the fourth chapter of this book.

Deepening social inequalities feature among the undesirable social consequences of human enhancement. In this book's fifth chapter, Jana Tomašovičová examines whether the principle of equality of opportunity can be considered to be a sufficient criterion for judging equality in a given situation. She argues that John Rawls's compensatory measures of equal opportunity do not sufficiently account for the diversity of human existence and therefore fail to ensure that no groups of people are excluded from the scope of justice and equality. She also analyses the capability approach and examines whether it is able to eliminate new forms of discrimination and exclusion that may arise as a result of cognitive enhancement and whether it incorporates the ability to cope with demands for the recognition of the equality of new and enhanced forms of life. Giving two primary reasons, Tomašovičová argues that the capability approach can be regarded as a more nuanced conceptual framework for thinking about equality in the context of human cognitive enhancement than what is provided by Rawls's theory of justice.

In the context of the latest gene-editing technologies, the author of the sixth chapter, Mariusz Pisarski, tries to map the movements at the boundary of scientific and artistic discourse. Similar to the numerous inspirations for human enhancement found in science fiction literature, the figurative language of Greek mythology provides several parallels with contemporary bioscientific discourse. Pisarski reflects on the technological possibilities of gene editing as a form of the contemporary rendition of the myth of Prometheus, who transcended established ethical norms in order to help humankind. Pisarski juxtaposes visions of the near future from cyberpunk and biopunk narratives with the contemporary discourse on the social, ethical, and economic implications of the application of biotechnology.

Bogumiła Suwara also deals with the issue of the interdisciplinary crossing of boundaries. In the seventh chapter of this book, she maps the problem of interdisciplinarity as an interface between two (and more) disciplines in contemporary literary science as well as in other sciences and in art. Using the example of digital literature and code poetry, she shows that the blurring of boundaries can lead to the integration of knowledge and methodological approaches as well as to the emergence of entirely new fields and subdisciplines such as critical code studies and visual studies. This can also take the form of a dispute, which is analysed herein using the three logics of interdisciplinarity (accountability, innovation, and ontology) formulated by Andrew Barry and Georgina Born. In Suwara's view, projects that seek new strategies of production are gradually pushing the pursuit of a distinct discursive autonomy into the background and primarily pay attention to the diffuseness of boundaries.

The authors of the chapters show that the shifts and blurring of boundaries that are currently taking place, particularly against the backdrop of changes in the life sciences, represent key moments for philosophical reflection. The challenge for contemporary philosophy is to enter into an intense dialogue with modern science, even while research processes are still underway. The shifting of boundaries ultimately forms a part of these boundaries' definition; upon the basis of a rationally guided discussion, these shifts can be guided and corrected so as to avoid any irreversible damage.

This book is dedicated to the memory of Juraj.

Jana Tomašovičová

Chapter 1

Robogenesis: The Automation of Human Reproduction

Juraj Odorčák

Abstract: Today artificial intelligence is being used experimentally to predict embryo viability. Some research teams have been also looking into the idea of the automation and robotization of other procedures that are frequently used in assisted reproduction. For a long time, other groups of researchers have been working on biotechnological devices that could ensure ectogenic gestation (ectogenesis). It is therefore imaginable that a synergistic combination of these and other reproductive technologies could one day lead to a complex procedure that would enable fully automated external human reproduction (herein referred to as robogenesis). This chapter aims to introduce the basic conceptual and ontological aspects of robogenesis. The first part is dedicated to defining robogenesis and situating it within the third wave of the reproductive revolution. The second part focuses on the analysis of theoretical, technological, and narrative assumptions about robogenesis. The final part presents arguments in favour of practising robogenesis. I will argue that a complete automation of human reproduction may be acceptable from the perspectives of humanism, transhumanism, and posthumanism, albeit for diametrically different ontological reasons.
Keywords: Reproduction, reproductive revolution, artificial intelligence, automation, robotization, ectogenesis, ontology, humanism, transhumanism, posthumanism.

Introduction

Ideas about diverse technological interventions in human reproduction have been part of various literary, philosophical, and ethical analyses for quite some time (e.g. Haldane 1924; Russell 1924). Some technological interventions in human reproduction have also been the subject of long-standing scholarly disputes (e.g. Deech and Smajdor 2007; Ross and Moll 2020). Such analyses usually focus on the topic of technological reproduction of some aspects of human reproduction (e.g. in vitro fertilization); however, a gradual change of parts can also lead to a change of the whole. Therefore, perhaps the most curious idea about technological modification of human reproduction is a vision of a complete technological reproduction of human reproduction. Such a hypothetical form of human reproduction will be herein referred to as robogenesis[2] (for an explanation of the term, see Section 1). We are still far away from achieving robogenesis; nevertheless, even today some research teams are working on experimental systems that could

[2] The term "robogenesis" is taken from the title of a science fiction work by D. H. Wilson (2014). Wilson uses the term to refer to the complicated genesis of robots. In this chapter, robogenesis refers to the complicated genesis of humans who would come into existence with the use of robots (robotic technologies, automation, and AI).

automate and robotize procedures related to assisted reproduction and gestation (Meseguer et al. 2012; Partridge et al. 2017; Varghese and Charalampos 2019; Casciani et al. 2021). Certainly, it is difficult to predict whether these experimental approaches will ever make it into clinical practice; however, it is not impossible to imagine a future where a synergistic combination of advanced reproductive technologies could enable fully automated external human reproduction (robogenesis). The strange character of robogenesis is related to unusual technologies as well as to possible extraordinary effects on the subject of robogenesis. It is precisely this subject which is the prime target of this chapter.

An analysis of the implications of a hypothetical technology requires conditional reasoning; therefore, the examination of robogenesis will proceed in the following manner: Section 1 outlines the definitional and situational background of robogenesis. It will point out how the conceptual assumptions about robogenesis relate to the direction of the ongoing reproductive revolution. The subsequent parts (Sections 2 and 3) analyse the science fiction and scientific assumptions about the direction of the reproductive revolution and robogenesis. A combination of visionary (literary) and technical (scientific) sources are used to identify both explicit and implicit assumptions associated with ideas about the radical transformation of human reproduction. Section 4 then explains the basic arguments in favour of practising robogenesis, which are related to different perceptions of the value of the origin of the human being itself. Indeed, it is the diverse understanding of the nature of the human being that is the main theme of this chapter. For this reason, I will move away from the bioethical aspects of robogenesis and instead focus on its ontological presuppositions and implications. I will argue that different ontological assumptions about humans (humanism, transhumanism, and posthumanism) can in fact lead to a positive evaluation of robogenesis (Section 5). From a humanist perspective, the main reason for applying robogenesis may be the desire to protect humanity's exceptional ontological status. For transhumanism, the primary reason for applying robogenesis may depend on the legitimization of the idea that there is a need to change the current status of human beings and enhance the human species. Posthumanism may see robogenesis as further evidence of the fundamental ontological openness of human beings. In the conclusion, I will argue that this ontological multiplicity of robogenesis implies a new argument about the inadequacy of the usual dichotomy between "natural" and "unnatural" (i.e. artificial) human reproduction.

1. The third wave of the reproductive revolution

The previous century was a period of intense scientific research in the field of reproductive biology, and research-based knowledge gradually led to the development of diverse reproductive technologies that modified biological aspects of the reproductive process of organisms (Frith 2012, 766). Organisms are time-limited entities; therefore, reproduction (and the potential modification of reproduction) is quite important for them in order to maintain the continuity (and potential change in continuity) of further life. Humans are also organisms, and thus it is not at all surprising that new insights in reproductive biology have led to the relatively rapid and widespread application of reproductive technologies to modify human reproduction. For this reason, some authors began to refer to this change as a "new revolution" (Lamm 1970). For example, John MacInnes and Julio Pérez Díaz (2009) have suggested that the rise of modernity is directly related to the economic and political transformation of human societies along with a revolution in human reproduction (henceforth referred to as the "reproductive revolution"). Admittedly, the exact theoretical boundaries of this revolution are disputed, as there are disagreements concerning the different views on the demographic, social, economic, historical, legal, and other definitions of the concept of the reproductive revolution (Robey et al. 1992; Nowak 2007; Daar 2017). From a technical perspective, however, the reproductive revolution can be defined through various ways of applying reproductive technologies.[3]

In the first wave of the reproductive revolution, reproductive technologies were typically used to temporarily prevent human reproduction (e.g. hormonal contraception). The second wave of the reproductive revolution was characterized by the application of new reproductive technologies that enabled people to reproduce when they otherwise – for a variety of reasons – would not have been able to do so (e.g. assisted reproduction). Both waves initially encountered a degree of misunderstanding (e.g. the derisive phrase "test-tube baby"), but the relevant reproductive technologies are now commonly used in many countries to modify human reproduction in various ways.[4] For this reason, the empirical results of

[3] In this chapter, the term "reproductive revolution" is understood only in the basic sense. This means that the reproductive revolution is a concept that approximately describes changes in human reproductive capabilities. A critique of the theoretical assumptions and possible normative implications of this concept is not the focus of this chapter; the reproductive revolution is used here mainly as a heuristic tool for determining different ways of applying reproductive technologies. The same rationale applies to the definitional demarcation of the boundaries of the different waves of the reproductive revolution.

[4] For example, in vitro fertilization (IVF) is now legal in all countries around the world. For a long time, Costa Rica had been the only country that did not allow it; however, since 2017 IVF has been allowed there following a Costa Rican Supreme Court ruling that invalidated a previous regulation banning it (Valerio et al. 2017, 366). The consideration of the case of the legality of IVF in Costa Rica shows that

these aspects of the reproductive revolution have also been relatively well documented (Daar 2017, 11–20); however, it is also true that the evaluations of some of the assumptions and implications of the application of these technologies are still subject to a relatively intense scholarly debate (Benagiano and Maurizio 2009; LeMoncheck 2020). This chapter will not outline a summary, interpretation, or ethical assessment of the application of known reproductive technologies of the previous waves of the reproductive revolution. Instead of the past or present, this chapter focuses on the possible future of human reproduction.

The third wave of the reproductive revolution will certainly depend in some way on a modification of the previous approaches of the reproductive revolution. Previous applications of reproductive technologies have often led to securing desired outcomes, but in some cases the use of these approaches has been associated with certain shortcomings and problematic consequences (Prudil and Pilka 2002; Pennigs and De Wert 2003, 397; Chatzinikolaou 2010). It is therefore likely that the third wave of the reproductive revolution will seek to avoid the shortcomings of the application of reproductive technologies used to date. One of the reasons for the failed application of reproductive technologies is the complexity of the biology of human reproduction and the ambivalence of human psychological reactions to technology. To put it another way, one of the problems with the application of reproductive technologies are human beings themselves. This is the case for two reasons. Humans make decisions about the application of reproductive technologies; however, because of the limits of their decision-making abilities, they sometimes make the wrong decisions. In other cases, people may make the right decisions, but the actual limits of human physiology may not allow these decisions to be implemented. The solution offered to these shortcomings is relatively simple but also controversial for that very reason. It consists of eliminating the cause of the shortcomings, i.e. limiting the (failing) human factor. If one is only concerned with the outcome, the third wave of the reproductive revolution could be directed towards the implementation of innovative reproductive technologies that would completely eliminate the shortcomings of human decision-making processes and fully externalize the biological aspects of human reproduction.

Such advanced reproductive technologies do not yet exist, so currently one can only speak of a hypothetical third wave of the reproductive revolution. On the

---

the reproductive revolution does not proceed in just one direction, since from 1995 to 2000 IVF had been legal and practised there (2017, 366). It should also be noted that the legality of a reproductive technology does not necessarily imply its availability or social acceptability and vice versa. Aspects of the reproductive revolution are subject to change, and these changes are the prime focus of this chapter.

other hand, some technologies that are used for similar purposes in other areas of life are already being considered for application in support of human reproduction. One of the current topics in reproductive medicine is testing the introduction of artificial intelligence (AI) (i.e. the automation of decision-making processes) into some procedures related to assisted reproduction (Varghese and Charalampos 2019; Zaninovic and Rosenwaks 2020; Casciani et al. 2021). It is also a fact that many research institutions are in the process of developing and refining procedures that would ensure the earliest and best possible extracorporeal development of foetuses (Partridge et al. 2017; Bulletti and Simon 2019; Segers 2021). These technologies are now primarily used to modify animal reproduction (Wilkinson and Di Stefano 2020, 43). If it turns out that these experimental procedures are also effective for human reproduction, then a combination of the two (i.e. fully automated and external human reproduction) is quite conceivable.

It is not at all difficult to suppose that this form of reproduction may cause a certain degree of caution and feelings of alienation or dismay. This is why I will refer to this hypothetical reproductive technology that would combine reproductive technologies with robotics, AI, and other advanced technologies by using the strange-sounding neologism "robogenesis". There are four reasons for making this unusual terminological decision. Firstly, the exotic tinge to the term "robogenesis" is meant to suggest quite bluntly that the topic of this chapter is not the current reproductive technologies of the first and second waves of the reproductive revolution but rather the potential reproductive technologies of the next possible reproductive revolution. And this includes the arguments for and against these technologies.[5] Secondly, the technological aspect of robogenesis is meant to refer quite explicitly to the idea of a possible integration of automation technologies (AI, robots)[6] with reproductive technologies. Thirdly, robogenesis is a combined term which in its first part refers to a technology but which in its second part refers to a process of creation and to an origin (genesis). This neologism is

[5] The thematic focus of this chapter is on the synergy of individual reproductive technologies, but the assessment of the outcome of this synergy (robogenesis) may have no bearing whatsoever on the assessment of the acceptability or otherwise of the assumptions of this synergy (i.e. AI in IVF). The neologism "robogenesis" is also intended to avoid the unnecessary argumentative contamination of those reproductive technologies that are already commonly used today for modifying human reproduction.

[6] The term robot (robotic) is used in its broader technological sense. It is also used here in a form that is closer to the original meaning given to it by Karel Čapek (1920). In R.U.R., Čapek associates robots with mechanical and biological properties (protoplasm). Furthermore, R.U.R. ends with an overlooked reference to a text from the Book of Genesis. For an interpretation of Čapek's definition of robots in the context of the preservation of the human species, see Odorčák (2020) and Odorčák and Bakošová (2021).

therefore also an allusion to the established term "ectogenesis".[7] Ectogenesis refers broadly to technologies that aim to ensure extracorporeal gestation (Singer and Wells [1985] 2006, 9–10); however, some consider this term to be inaccurate, as the technologies in question only provide external (*ecto-*) gestation (development) (Kingma and Fin 2020, 356). These authors also point to the ambiguity of the term "origin" (genesis, emergence/development). This ambivalence of genesis is also a key aspect of those technologies that would automate human reproduction (i.e. robogenesis).

In short, robogenesis is the personification of a futuristic vision on where the development of reproductive technologies may eventually lead and a representation of a future where "every step of the *reproductive cycle will be automated*" (Haroon 2021). In this chapter, robogenesis will therefore refer to a hypothetical reproductive technology that will effectively provide fully automated external human reproduction. This definition refers to the combination of two fundamental aspects of robogenesis: automated reproduction and external human reproduction. In the case of robogenesis, human reproduction would be carried out through advanced automation technologies that would autonomously carry out the activities necessary for embryo formation and viability. At the same time, this technology would ensure gestation of the embryo (and the foetus) in an automated extracorporeal manner. Simply put, robogenesis combines the automation of assisted reproduction with the automation of gestation (externalization; ectogenesis). The automation (Casciani et al. 2021), robotization (Sroga et al. 2008), or mechanization (Meseguer et al. 2012) of assisted reproduction is a relatively new idea. On the other hand, visions of externalizing human reproduction have been part of the popular discourse on reproductive technologies for a relatively long time. Some have even suggested that the discourse on reproductive technologies itself is determined by reactions to popular science-fiction ideas about the possibilities of externalizing gestation (Aristarkhova 2005, 44). In the following sections, I will delineate the basic narratives about the externalization of human gestation and then turn to contemporary proposals for automating human reproduction.

2. The automation of gestation

Probably the most famous work that discusses the possibilities of an external form of human reproduction is Aldous Huxley's dystopian novel *Brave New World*

[7] Ectogenesis is a term that is a compound of the Greek *ecto* (external) and *genesis* (origin) (Vallverdú and Boix 2019, 107).

([1932] 2018). In this novel, the externalization of human gestation is portrayed as one of the conditions that enable the political, social, psychological, and technical manipulation of individuals through a global authoritarian state system. Huxley himself understood his "negative utopia" (1963, 232) as a critique of the "horrors of the Wellsian Utopia" (Smith 1969, 348). The place (*topoi*) and actual content of this dystopia are debated, but one common interpretation is that it is a critique of political projects in the United Kingdom in the interwar period (Von Miese 1944, 110). In addition to its political aspects, Huxley's novel also reflected on the philosophical and scientific assumptions of modern society. It is therefore not surprising that Huxley did not address the topic of ectogenesis for the first time in *Brave New World*. Huxley's original depiction of the externalization of human gestation can actually be found in his first novel, *Crome Yellow* ([1921] 2018). In this novel, Huxley introduces ectogenesis through the views of a character named Mr Scogan – who, according to some scholars, personifies Bertrand Russell[8] (Montgomery 1974; Moran 1984). Mr Scogan proposes the "industrialization" of pregnancy as one of the possible solutions to the shortcomings of human society (Tripp 2015, 32). Although Huxley satirically points out the absurd consequences of Mr Scogan's reductive philosophy,[9] the augmentation of the possibilities of human reproduction through "gravid bottles" ([1921] 2018, 47) is also presented as a hope[10] that "biologists may educate society by enabling it to use science wisely" (Lewicky 2008, 213). This narrative[11] portrays the externalization of reproduction as a tool that enables the development of some hidden human

[8] Huxley met Russell while studying at the University of Oxford. His attitude towards Russell was initially critical, as he disapproved of the closed nature of the community that brought together Russell and other prominent intellectuals of the United Kingdom (the "Garsingtonians"). Huxley faulted this community for preferring to carry out its own slogan of "saving England" as a trip "to a kind of rustic Bloomsbury to avoid reality and live the life of the mind" (Meckier 2003, 87). After the First World War, Huxley's relationship with Russell changed, probably under the influence of Huxley's attendance at lectures at the Cambridge Heretics' Society (Saunders 2019, 35). In this period, Huxley's work contains explicit references to theories put forward by Russell himself (Marovitz 2003, 145).

[9] Huxley described Mr Scogan as follows: "In appearance Mr Scogan was like one of those extinct bird-lizards of the Tertiary... His movements were marked by the lizard's disconcertingly abrupt clockwork speed; his speech was thin, fluty, and dry... Mr Scogan might look like an extinct saurian" (2018, 21).

[10] Huxley described this hope in the following way: "An impersonal generation will take the place of Nature's hideous system. In vast state incubators, rows upon rows of gravid bottles will supply the world with the population it requires. The family system will disappear; society, sapped at its very base, will have to find new foundations; and Eros, beautifully and irresponsibly free, will flit like a gay butterfly from flower to flower through a sunlit world. 'It sounds lovely', said Anne. 'The distant future always does'" ([1921] 2018, 47).

[11] Some authors (e.g. Thody 1973, 50) have suggested that *Brave New World* is merely a literary adaptation of Russell's *The Scientific Outlook* ([1931] 2001). In his review of *Brave New World*, Russell praised Huxley's literary talent (Russell [1932] 1997, 210) but also argued that popular opposition to ectogenesis stems mainly from man's desperate longing for the illusion of free will ([1932] 1997, 212).

characteristics. Ectogenesis and reproductive technologies only magnify[12] and approximate the essential characteristics of humanity.

Optimism about the science and accomplishment of ectogenesis can also be found in another classic work which is usually considered the inspiration for visions of a new reproductive world. J. B. Haldane's *Daedalus; or, Science and the Future* (1924) is considered to be the first work to introduce the term "ectogenesis" into wider public discourse (Ferreira 2017, 136). Haldane attempted to estimate the most likely development of scientific progress and its predicted impact on human society (Berenbaum 2012, 123). *Daedalus; or, Science and the Future* is therefore constructed upon the basis of a retrospective narrative approach, i.e. from a future perspective on the present (or past). Haldane places ordinary human reproduction, among other things, in the past. He predicts ectogenesis and cloning as the most common and secure forms of future human reproduction (Haldane 1924, 63). However, Haldane also links these reproductive technologies to the selection of "ancestors for the next generation based on their genetic superiority"[13] (Jeffreys 2001, 140). In Haldane's prognosis, ectogenesis itself is defined both as an "ordinary" reproductive technology as well as a conceptual tool that allows us to depict the mutable nature of the human species. Haldane presents the transformation of humans and their reproduction as a possibility and as an existential necessity. In *Daedalus; or, Science and the Future*, reproductive technologies are proposed to address human infertility as measures to overcome the fragility and contingency of human reproduction (Haldane 1924, 63). Within this narrative, the technologization of reproduction is thus primarily an expression of the rational pursuit of (species) self-preservation. For Haldane, ectogenesis and reproductive technologies are primarily ways to overcome the unsuitable biological limits of humans and are therefore tools for the improvement of humanity.[14]

A more extravagant approach to ectogenesis (and technology in general) can be found in J. D. Bernal's provocatively entitled *The World, The Flesh and*

---

[12] This is why Huxley portrays ectogenesis and reproductive technologies differently in different works.

[13] One of the first responses to *Daedalus; or, Science and the Future* was a critical essay by Russell. In *Icarus, or the Future of Science* (1924), Russell criticizes eugenics because of the risk of its misuse by state or private actors. In a joint review of both works, an anonymous contributor for *Nature* expressed his "hope that Mr Haldane's booklet will not lack readers" (Anonymous 1924, 740) and described Russell as a writer who "dislikes present-day Western civilization" (1924, 742).

[14] Haldane presents this view explicitly in his essay entitled *The Last Judgment* ([1927] 2017). This essay discusses the future end of the Earth and the extinction of humanity. Haldane, from the perspective of a narrator from the future, warns against the use of technology to stop evolution. According to Haldane, the fundamental mistakes of Earthlings were (or are) the application of technology to the perfection of human subjective qualities (happiness) and the attempt to technologically preserve (the limits of) humanity ([1927] 2017, 292).

the Devil ([1929] 2017). Unlike Huxley and Haldane, Bernal posits that science and technology will not only define (Huxley) and improve (Haldane) humanity but also radically change it. Bernal starts from the idea that the enemy of the rational soul (i.e. scientific knowledge and creative thinking) is the tendency of humans to limit science and technology to theories and applications that are seen only from the perspective of humans as individuals ([1929] 2017, 42). He points out that science is both valuable to humans as well as in itself ([1929] 2017, 42). This is also true of the change of reproduction. In this approach, alongside other technologies, ectogenesis is portrayed as one of the steps that aim at the thorough destruction of all the usual biological determinations of human beings. The externalization of human reproduction, which Bernal characterizes through a vision of modifying "the germ plasm or the living structure of the body, or both together" (Bernal [1929] 2017, 35), however, does not merely aim at liberating humans from the "requirements" of conventional reproduction. Bernal sees the real purpose of externalizing reproduction in the ontological liberation of creativity itself.[15] One form of creativity's liberation may be the emergence of a new species that uses both habitual and seemingly unusual ways (ectogenesis) of reproducing and sustaining its own life. Other forms of ontological personification of creativity can take forms that far exceed our most fanciful expectations and ideas of what is human, reproduction, species, nature, science, technology, and their interrelationships. In this sense, ectogenesis is just one conceptual tool that depicts the relativity of humans and the inadequacy of species-oriented conceptions of humanity. Bernal's approach to ectogenesis is thus more radical than Huxley's and Haldane's visions. He proposes reasons for the modification of human reproduction that are (perhaps paradoxically) independent of the biological assumptions, demands, and sentiments of ordinary human individuals. In this narrative, reproductive technologies are an example that points to the need to transform and transcend humanity itself.

It is therefore not surprising that these science fiction visions of human (reproductive) change have sparked a rather large wave of both specialized and general interest in the direction of reproductive technologies. Haldane's prediction caused a sensation and had an exceptional readership (Adams 2000, 462). However, speculation about ectogenesis also generated determined public opposition to the possible application of such reproductive technologies right from the start (Ball 2011, 202). The idea of externalizing reproduction led to a long list of highly

[15] Some authors therefore interpret Bernal's conception of creativity as a science-fiction personification of *Eros* (Hassan 1979, 130; Ferreira 2011, 122).

critical, albeit variously motivated, responses (e.g. Ludovici 1924; Brittain 1929). It has also led to ectogenesis becoming a fairly common dramatic trope in many works of popular culture. Not only did the term "ectogenesis" become a part of the lexicon, but so did problematic expressions such as "test-tube baby",[16] "artificial womb",[17] and even "abnormal reproduction".[18] Many therefore assume that it was these visions that led to the public's initial reserved stance also towards other reproductive technologies (e.g. IVF) (Franklin 2013, 75; Ball 2013, 339). The problem lay in the public's persistent perceptions of the intertwining of reproductive technologies with specific political and national goals. Some authors have suggested that it is precisely such problematic perceptions that have led to the emergence of the non-state reproductive healthcare provider model in some countries (i.e. the United Kingdom) (Ferber et al. 2020, 257–258).

Beyond the straightforward political critique, these visions of ectogenesis also foreshadowed the problem of correctly depicting the meaning of the application of science and technology. The trope of the "mad scientist" producing "decanted monsters" may be more of a topic of fringe literature and dubious Internet forums in the present day, but real-world depictions of the results of scientific teams currently working on ectogenesis still lead to repeated references to some "brave new worlds" (Derbyshire 2019, 1; Zimmer 2021, 29). The enduring allure of the symbolism of these (now quite old) science fiction visions of reproduction is thus probably also related to the compelling portrayal of the idea of the human world as a purely scientific and technical problem. In such a depiction, the world is just an experimental factory and a precisely organized laboratory.[19] This causes interest among recipients and dismay among critics. In other words, the appeal of these visions of ectogenesis lies in the pregnant expression of the simple and thus precisely controversial idea that reproduction is only production (Lucke 2019, 345). And if it is true that in this approach to reproduction human children are not

[16] France Winddance Twine assumes that the term "test-tube baby" had been introduced by Huxley in *Brave New World* (Twine 2015, 6); however, other authors point out that the term had been used much earlier in connection with artificial insemination and embryo culture (Wilson 2011, 53). Moreover, in addition to being misleading, the term is also inaccurate; it would be "more correct" to speak of embryos from a Petri dish.

[17] There is currently an ongoing controversy over the use of this term. Elselijn Kingma and Suki Finn suggest that it may express the power and political implications of the technology (2020, 361). Others argue in favour of using the term due to the fact that it is quite well established and easily understood (Romanis et al. 2020, 1).

[18] This curious expression was pointed out by George Annas in his analysis of the various names for assisted reproduction (1984, 1415).

[19] In this context, Federico Neresini speaks of a "laboratization of the world" (2011, 67). He defines laboratization as practice where "science is constantly engaged in an attempt to transform the natural/social environment according to its needs" (2011, 67).

begotten but made (Kigozi 2018, 42), then the human itself is also understood here as a peculiar product. In this sense, these visions and their critiques express understandable political concerns as well as various ontological assumptions about humans and the world. The dispute between the proponents and opponents of these technologies then mainly lies in which of the geneses and ontologies they consider to be more existentially stable. Other things being equal, opponents usually assume that the "providence of" nature (in the broad sense) is a more stable phenomenon, whereas proponents usually assume that the better option is the prudent use of science and technology. In both cases, the human being is a kind of creation in a sense, but the question is who is the better creator: nature mediated to humans or human culture? Having said that, in this binary perception there is a fundamental flaw to these visions and their critics. From today's point of view, they are based on presuppositions about anachronistic modes of creation or production. Therefore, they also start from an incomplete number of possible geneses and an incomplete set of ontologies of human origins.

3. The automation of reproduction

Even the most radical vision of ectogenesis assumed that humankind's future would be the result of its creative choices. Bernal did argue that the biology of human reproduction would change (or, paradoxically, be lost; [1929] 2017, 41) but only in order to preserve the human spirit ([1929] 2017, 42). Even in this fantastical approach, human modification (nature) is proposed mainly in order to develop human rationality and creativity (culture) in unbridled (and possibly grotesque) forms. In this respect, however, these science fiction notions of the modification of human reproduction are proportionally outdated, since even such exotic visions (be they perceived as utopias or dystopias) have failed to estimate with complete accuracy the present possibilities for the transformation of humans and the world. It seems no longer obvious that it is not only (in Bernal's vocabulary) the flesh that is automatable but the soul as well. This "soul" then does not need to be human at all, and this is what is now also considered to be its fundamental advantage.

The ongoing revolution in automation is therefore based on requirements that are similar to the reasons in favour of the application of the technical improvement or replacement of (parts of) human physiology. If it seems reasonable to modify, improve, or replace the human body, then it may seem similarly reasonable to automate human decision-making processes. The current automation revolution, however, is not a concept that is completely distinct and compact. In some cases, it is associated with specific technologies (Schlick 2012) or production processes of the "fourth industrial revolution" (Kopacek 2019). In other cases, this revolution is again considered to be a process that takes place at the technological as well as at the organizational, global, political, and social levels (Helbing 2015). In addition to the technological and organizational impact of this automation, the exact timing of the revolution in question is also unclear. Some locate the beginning of the automation revolution in the last decade (Scholl and Hanson 2020), whereas others place it at the turn of the millennium (Smith 2018, 13); some historians even locate the birth of this trend in a much earlier time (Luckhurst 2014, 318). Setting aside these definitional ambiguities, in basic terms the current automation revolution can be simply seen as an effort to automate those aspects of the world that until recently were considered difficult to automate or were seen as completely non-automatable. The automation revolution is thus aimed at effectively automating the diverse cognitive abilities of humans (the "soul"), which is why it is sometimes called "intelligent automation" (Wirtz et al. 2021, 38) and even the automation of everything (Kuru and Yetgin 2019, 41395). This in turn raises the natural question of whether reproduction should be included in the set of this "everything".

Not surprisingly, however, assisted reproduction has long been an area that has resisted the demands of applying such automation approaches (Gupta 2020). Reproduction is related to existence, so a certain caution and conservative approach to the application of new technologies is always a natural and necessary part of the responsible practice of assisted reproduction. The acceptability of assisted reproduction is contingent on confidence in its safety and is fundamentally dependent on there being hope in its success. An essential advantage of automation is its efficiency. Recently, some critics have therefore begun to point to assisted reproduction's fundamental lag behind ongoing technological advances in the life sciences, biology, biomedicine, and even standards of laboratory work (Varghese and Charalampos 2019, 848). In biology, machine learning technologies are now proposed for things such as limiting the human inability to predict the behaviour of biological systems (Carbonell et al. 2019, 1474). In biomedicine, the use of technologies such as automatic control, autonomous execution, AI, and robotics is already a fairly established practice for improving the quality and efficiency of healthcare (Graur et al. 2010, 457; Pang et al. 2018, 251). Also, the automation of laboratory work is a long-term trend that has been underway for the last forty years (Holland and Davies 2020).[20] From this perspective, it is easy to understand why some critics have expressed astonishment at the current state of assisted reproduction. Nonetheless, assisted reproduction is an area where, for obvious reasons, there is little room for experimental approaches and the potential errors associated with them. Proponents of the application of automation in assisted reproduction therefore usually argue in favour of using already proven advanced technologies that would be sensitively adapted to the possibilities and needs of specific interventions in human reproduction (Casciani et al. 2021). Proponents of this idea often point to the need for a judicious consideration of the implications of changing standards in assisted reproduction (Kragh and Karstoft 2021). Outcomes are also a fundamental topic of debate about the future of assisted reproduction, as a growing number of actors in this debate assert that the proper automation of assisted reproduction could dramatically increase its success rate while significantly reducing the costs of application (Meseguer et al. 2012; Varghese and Charalampos 2019; Casciani et al. 2021). The implementation of robotics and AI is considered to be the most promising route to such an outcome.

Research into the use of robotic assistance technology for micromanipulation-assisted reproductive techniques is currently underway. Such technologies include robotic intracytoplasmic sperm injection, which uses the precision of robotic assistance to immobilize and then insert the sperm into the oocyte (Zhe et al. 2011, 2102). The advantages of this technology lie in minimizing the need for human intervention, high reproducibility, and a corresponding success rate of insemination (Zhe et al. 2011, 2102). Other research teams describe the possibilities of the non-invasive robotic spatial manipulation of embryos (Huang et al. 2021) and robotic assistance in the vitrification and cryopreservation of embryos (Varghese and Charalampos 2019, 852). Robotic assistance technologies are also already being considered for optimizing gamete and embryo selection (Wang et al. 2019, 139). Robotics in this area would allow for the reduction of the risk of potential damage to gametes and embryos, as it would reduce the number of steps currently required for the assisted reproduction procedures in question. Consequently, this procedure could also lead to the greater standardization of embryo and gamete handling (Kragh and Karstoft 2021). Furthermore, research into the application of robotics in assisted reproduction is not only taking place at the level

[20] However, Ian Holland and Jamie Davis also point out that in all areas of life sciences there is some resistance to automating laboratory work (2020). A more cautious approach to automation is particularly visible at academic research institutions; however, the reason for this reluctance may not lie in the resistance of the researchers themselves but rather in the nature of academic research funding (2020).

of the "microcosm". Robotic assistive systems are now also successfully used in surgical and other medical procedures that are applied in the context of the human reproductive system (Jayakumaran et al. 2017). The indisputable advantage of such interventions in the reproductive system is greater precision and less invasiveness (Sroga et al. 2008, 1308). Some even suggest that many other aspects of assisted reproduction could be robotized in a similar "micro-" or "macro-" manner (Meseguer et al. 2012). Essentially all assisted reproduction procedures that are characterized by repetition, while also requiring high precision, controllability, and safety, are a fundamental objective of such robotization.

Having said that, assisted reproduction certainly requires accurate and safe procedures as well as proper evaluations of those procedures. The second part of the proposals for automating assisted reproduction therefore aims at streamlining the decision-making processes of the professionals that are involved. One of the essential prerequisites for successful assisted reproduction is the identification of an embryo that is suitable for transfer to the uterus for subsequent gestation.[21] At the present time, the assessment of embryo viability is mainly based on the expertise of embryologists (Lundin and Ahlström 2015, 460). Embryologists determine (or estimate) the predicted viability of the embryo based on observations of morphological and other features of particular embryos (Lundin and Ahlström 2015, 460). Gaining proficiency in assessing viability is not a trivial matter, as mastering this skill requires relatively long training and extensive experience in embryo assessment (Khosravi et al. 2019, 21). This activity is also relatively costly. There are also some differences among experts in assessing embryo viability, given that there are different viability assessment protocols for this activity (Nasiri and Eftekhari-Yazdi 2015), and therefore embryo selection for assisted reproduction depends on subjective human decisions. This subjectivity in decision-making processes could be reduced or fully eliminated in the future by the proper use of AI. In some cases, algorithms that use deep machine learning methods can already estimate embryo viability just as well as some embryologists (Khosravi et al. 2019, 21). The reason is quite simple. These algorithms can access a database of viable embryo examples that far surpasses the experience of any embryologist. These databases are also continuously expanding, so it can be assumed that the capabilities of the algorithms to determine embryo viability will continue to improve. For this reason, proponents of the application of AI in assisted reproduction

[21] Currently, single-embryo transfer is preferred in assisted reproduction. Single-embryo transfer aims to avoid the negative consequences that are naturally associated with multiple pregnancy. Assessing embryo viability therefore plays an even more important role today for the success of assisted reproduction.

suggest that the use of AI could lead to an improved assessment of gamete quality and embryo viability (Zaninovic and Rosenwaks 2020, 914) as well as a more efficient estimation of gestational success (Miyagi et al. 2019) and even a prediction of the success of assisted reproduction as such (Goyal et al. 2020). This optimism is based on the notion that AI has the unique potential to incorporate the biological and social dissimilarities of individuals at the level of fertility (Trolice et al. 2021). In other words, it is assumed that the analysis of large amounts of medical, genetic, and social data using AI can lead to the discovery of new solutions that would enhance human reproductive capabilities.[22] From a theoretical perspective, AI is thus applicable to all areas of assisted reproduction that require challenging decision-making and creative processes; however, even if the applicability of AI in assisted reproduction is definitively confirmed, the practical question of how to specifically incorporate AI into decision-making processes concerning assisted reproduction remains open.

A similar problem can be seen in all other areas where the use of AI is proposed. In general terms, three basic ways of incorporating AI into decision-making processes can be determined.[23] The first way of incorporating AI into decision-making processes uses technology to merely inform human decision-making (Ouyang and Jiao 2021, 2). In this case, AI only provides the informational basis that enables a human to make a certain decision, take certain action, or perform a certain activity. This sort of AI has an advisory and informational function (Guzman 2016, 69), and the main actor is the human. The second way of incorporating AI into decision-making processes uses AI to monitor and possibly change the results of some human decision-making processes. In this approach, AI provides information and ensures the limitation of possible incorrect human decisions (Hoc 2007, 283). The second form of AI incorporation is therefore mainly used in activities and situations where there is a risk of some level of fatigue and loss of human attention, which is why it is sometimes referred to as "peer AI" (Gromyko et al. 2017, 238). In this case, the human and AI are co-actors in a certain activity. The final way of applying AI in decision-making processes

[22] However, such optimism about AI is not shared by all participants in this debate (Trolice et al. 2021). The problem with the current application of AI in assisted reproduction lies mainly in the fact that there are still no studies that sufficiently confirm the suitability of such technologies for the clinical practice of assisted reproduction (Casciani et al. 2021).

[23] The definition of this distribution is built upon a theoretical model proposed by Fan Ouyang and Pengcheng Jiao for paradigms for the use of AI in education (2021). Ouyang and Jiao argue that education is moving towards a reflexive use of AI (2021, 1). This model is applied here to the problem of AI use in assisted reproduction, but this does not mean to imply that the uses of AI in assisted reproduction would necessarily be the same.

is based on the assumption that applying the results of human decision-making processes is unnecessary and harmful in certain cases. The problem with human decisions may be that human decision-making capabilities have certain limits that lead to the unpredictability of these decisions. Indeed, it is precisely the unpredictability and subjective nature of human decisions that can in some cases lead to negative consequences. Therefore, this way of incorporating AI into decision-making processes uses AI to directly manage the human decision-making processes (Ouyang and Jiao 2021, 2). AI plays a major active role in this approach as human decision-making processes are completely replaced by automated management.

From a theoretical point of view, there are therefore three ways of applying AI in assisted reproduction. Based on an analysis of large amounts of data and examples, AI can provide experts with valuable information that will increase the likelihood of a decision being correct. For instance, AI could provide information about the likely viability of an embryo; an embryologist could then use this information and his own expertise to make a definitive decision about the quality of that embryo. As noted above, such an application of AI in assisted reproduction is being used in experimental form today. However, some authors hypothesize that AI will be able to assess embryo viability even more accurately, reliably, and quickly than any human embryologist (VerMilyea et al. 2020, 772). If this ability of AI is indeed confirmed in non-experimental settings, then it is very easy to imagine AI being incorporated into the direct supervision of embryo quality assessment as well. In this case, AI would prevent the embryologist from incorrectly assessing embryo viability. Such an incorporation of AI would also be useful for making the training of new professionals (collaboration) more effective, thus increasing the efficiency of the whole assisted reproduction process. Similar ways of applying AI can be imagined in other areas of assisted reproduction, although it is also clear that there will be fundamental reservations about these applications of AI as well, based on concerns regarding the safety of the technologies and procedures involved (Kragh and Karstoft 2021). Safety is the core reason for a third possible way of applying AI in assisted reproduction. An assessment of embryo viability is carried out in cases of assisted reproduction to ensure the highest probability of successful gestation. If embryo viability provides the best estimate for the probability of successful gestation, and AI can identify the most viable embryo, then it is not at all obvious why that particular embryo should not be used for a particular embryo transfer, gestation, and assisted reproduction. If AI is making the right decision in this case, then it makes sense to make that decision and only that decision. Thus, from a theoretical point of view, a third way of incorporating AI into assisted reproduction is quite conceivable. AI would decide on the relevant assisted reproduction procedures and thus the outcome of human reproduction.

Of course, none of the participants in today's debate about the introduction of AI into assisted reproduction explicitly argue in favour of removing experts from the decision-making and creative processes related to assisted reproduction. The debate is rather usually framed by the familiar slogan that "humans should stay in the loop" (Johnson 2008, 535). On the other hand, some authors are already using examples in this context that refer to the automation that takes place in the automotive and other industries (Casciani et al. 2021), which illustrates that the future role of humans in the automation loop can be imagined in very different ways. The current scholarly debate is largely concerned with the technical, safety, and medical aspects of the first and second ways of incorporating automation into assisted reproduction. The same is true for the ethical evaluation of these innovative options.[24] In other words, the discussion today is mainly focused on aspects of the partial introduction of automation into reproduction, as currently there are only technologies that allow the partial robotization and partial introduction of AI into assisted reproduction. The same applies to ectogenesis; however, the partial automation of human reproduction is not the present focus. This chapter deals with robogenesis, which is a hypothetical technology that would fully combine all the proposed ways of automating human reproduction. The fundamental question then is what reasons might lead someone to accept the application of fully automated external human reproduction.

4. Arguments in favour of robogenesis

The answer to this question may be very simple. Robogenesis would mean one more reproductive option. If we care about increasing reproductive options, then we could accept the application of robogenesis. Furthermore, the demand for more options could be justified in two different ways. More options are good because they increase the number of good choices. In this case, the justification is directed at increasing the total sum of good choices. However, a larger number of choices can also be good because it limits the necessity of choosing the wrong choice. In

[24] At the time of writing this chapter, there was only a single (preprint) study (Afnan et al. 2021) that explicitly addressed the issue of the ethical evaluation of the (partial) introduction of AI into IVF. In this respect, the evolution of technology has outpaced the evolution of ethics. The authors of the above study point out many ethical issues related to AI and suggest reasonable ways to regulate the application of this technology to IVF (Afnan et al. 2021).

the latter case, the justification is more likely motivated by a proportional reduction in the number of bad choices ("evil"). A greater number of choices may help us gain something, while it also may help us avoid losing something. This is also true in the rationale for robogenesis.

The first way of justifying the application of robogenesis is based on an ideological approach that assumes that if we have the appropriate technological means (capabilities) to modify human biology according to our preferences, then we should use these means (technology).[25] However, this approach is only an expression of a more general principle that states that if we care about a certain outcome that is considered good from a general point of view, and we also have the necessary tools to achieve that outcome, then we should use those tools (options) to achieve that outcome. In the case of robogenesis, such a goal is further existence. Since further existence is usually considered a good in itself, technology that contributes (or can contribute) to that goal should not be considered bad (or should not be prohibited). Thus, a proponent of robogenesis would point out that this technology is merely a very advanced and unusual tool to provide for our normal efforts to match requirements with the reproductive possibilities. The requirements would be met by changing the possibilities of achieving them. Robogenesis would only increase the sum of opportunities to realize valuable wishes. Under certain circumstances, the justification for an application of robogenesis that does not operate solely with the concept of the fulfilment of desires or wishes is also conceivable. A second justification of robogenesis might assume that the application of this technology may be necessary in certain circumstances. Proponents of such a justification would argue that in some cases, robogenesis may be considered a necessary condition for the application of reproduction. The validity of such a justification, however, depends crucially on an explication of the circumstances upon the basis of which we would have grounds for making the use of robogenesis obligatory. The reasons for such an obligation may not only lie in some malevolent plans of totalitarian politicians (*Brave New World*) but may also be naturally related to the application of sound principles of individual freedom and responsibility for the future of humankind. The first reason for the obligation of robogenesis could be based on the well-known principle of procreative

[25] Similar arguments for the modification of human biology can be found in the current debate about the "enhancement of love". Proponents of the technological modification of love assume that the application of things such as pharmacological techniques could lead to more stable (or preferred) forms of human romantic cohabitation (Earp and Savulescu 2020).

beneficence (Savulescu 2001).[26] According to this principle, reproductive actors have a personal responsibility for the best outcome of their reproduction (Hotke 2014, 255). If robogenesis could provide a reproductive outcome that was better than other forms of conventional and assisted reproduction, then, based on the principle of procreative beneficence, reproductive actors would have an obligation to choose robogenesis for their reproduction. A second reason for the obligation of robogenesis could be based on an assessment of the goodness of group reproductive outcomes, i.e. the responsibility to save the human species. The principle of procreative beneficence would be replaced by the principle of preserving long-term human survival (Munévar 2014, 197). When considering the premises of safety, it would be sufficient to add the factual premises of the future destruction of the Earth as well as the complexities of normal human reproduction in an extra-terrestrial environment (e.g. gravity, radiation, and resources). In these circumstances, robogenesis could be portrayed as an essential condition that would prevent the extinction of humanity.[27] From this perspective, the obligation would be to use robogenesis (i.e. survival in space) as well as undertake research into robogenesis (i.e. the prevention of humanity's extinction). Robogenesis can thus be imagined as a "luxury" option that would fulfil the diverse wishes of variously motivated individuals as well as a vital necessity that would ensure the existence of some individuals or, indeed, humanity as a whole.

5. Robogenesis and humanism, transhumanism, and posthumanism

The acceptability of such arguments in favour of robogenesis depends on the justifications for the conceivable change in human reproductive capacities and on an overall understanding of that aspect of robogenesis which would be seen as its primary goal. Different theories about humans may therefore lead to different understandings of the purpose of applying robogenesis. One of the most prominent ideological concepts regarding humans is humanism. In a broad sense, humanism is a collective label for a group of philosophical theories that primarily focuses on defining the status of human beings and their relationship to other objects and subjects in the world (Setiya 2018, 457). For humanism, the most fundamental

[26] The principle of procreative beneficence is used here as an illustrative example to amplify some aspects of the potential application of robogenesis. The choice of this example does not mean to imply that this principle is irrefutable. It is beyond the scope of this chapter to delineate the advantages and disadvantages of procreative beneficence. For arguments for and against the principle of procreative beneficence, see the study by Andrew Hotke (2014).

[27] A similar argument concerning ectogenesis is presented by Matthew Edwards (2021). He argues that embryo space colonization technology has greater potential for the preservation of the human species than the usual proposals to colonize the galaxy through manned space missions (2021, 323).

issue is the determination of the proper value of human beings and the subsequent structure of human relations to the world. Humanist philosophies are primarily divided according to different perceptions of these relations. In some humanist philosophies, the central relation is the connection of humans to a transcendent being (religious humanism; De Gruchy 2018), while in others the fundamental relation is that of humans to nature (secular humanism; Felderhof 2011). Some humanist philosophies see the crucial relation of humans as being towards a particular community or to humanity itself (social humanism; Ellis 2012). Other philosophies of humanism see the most important relation of humans in connection to life and the meaning of existence (existentialist humanism; Melhi et al. 2020). All these philosophies build such a structure of diverse relations upon a common assumption about the particularity of human beings (Figdor 2021, 1546). Humanistic philosophies assume that humans and humanity are both special, unique, and irreplaceable in some way. Humanism considers humanity as an entity that has a specific status (exceptionalism). This status sets humanity apart from other entities in the world (uniqueness). At the same time, this identity of humanity (specialness, exceptionality) is the basis for the idea of humanity's irreplaceable position in the world. Different humanistic philosophies then privilege different essential characteristics that ensure the special status of the human being. The essential characteristics are usually derived from an understanding of a human's preferred relationship to the world (i.e. soul, rationality, and creativity).

All of these characteristics, however, in some way express the rather simple idea of a human's perfectibility: "What a Piece of Work is Man!" (Shakespeare [1600–1601] 2000, 85).[28] If this perfection does exist, it is fitting that it should continue to do so. For humanism, therefore, human reproduction is one of the fundamental aspects of its assumptions about human beings (Hafer 2020), since quite obviously, without human reproduction, exceptional human beings would not exist, and thus humanism itself would not be possible. Of course, the fundamental controversy in humanism is the question of the essence of humankind, so different forms of humanism may prefer those modes of human reproduction that, according to the assumptions of that particular humanism, preserve the proper essence of humans. On the other hand, if the humanisms in question are humanisms not only for the sake of reproducing humanism itself (ideology), but are primarily for the sake of seeking to preserve humankind, then reproductive

[28] Hamlet's exclamation is interpreted by some authors as Shakespeare's expression of the basic paradigm of humanism (Nowottny 1964, 63). Others point out that Hamlet's exclamation, as well as Shakespeare's relationship to humanism, is open to multiple interpretations (Garabedian 1996; Norman 2004, 1–2).

technologies that appropriately fulfil this goal should be acceptable to humanism.[29] This should probably also be true of robogenesis, since the operational mode of such technology is to increase the possibilities for human reproduction and thus increase the possibilities for human preservation. From a humanist point of view, both variants of the argument in favour of applying robogenesis might be acceptable under certain circumstances. In good times, robogenesis could simply mean an option pointing to human exceptionalism. In bad times, robogenesis would imply a condition that would protect the existence (exceptionalism) of the human being, the human species, and humanism itself. From the point of view of humanism, the general sense of applying robogenesis may be to highlight the exceptional ontology of human beings, which must be protected at all times. For humanism, even such an extravagant technology as robogenesis can serve as a tool to preserve the "masterpiece" or preserve the belief in the "masterpiece" that is the human being.

Humans can be characterized through beauty as well as through a certain misery, since the human species also exhibits some biological, psychological, and other deficiencies (Gehlen [1957] 1980; Tomašovičová 2021, 31). These deficiencies can be evaluated in two different ways, which delineate two other concepts of the status of human beings. Transhumanism is a school of thought that is based on the common notion that some human failings are the very reason for human exceptionalism.[30] Unlike humanism, transhumanism assumes that humans have yet to become some sort of "masterpiece". Transhumanism thus agrees with humanism on the idea of the existence of human exceptionalism (Cordeiro 2019, 70). For transhumanism, however, the exceptional status of human beings lies primarily in human ingenuity, which is principally aimed at overcoming all human shortcomings (Rähme 2021, 119). Since transhumanism views the overcoming of all human shortcomings as an essential human characteristic, all forms of overcoming shortcomings are seen as good (Clark 2013, 124). Some transhumanists also believe that, in certain circumstances, human enhancement can also lead to a fundamental transformation and overcoming of humanity and to the creation

---

[29] The conflict between different forms of humanism also lies in the important consideration of whether a particular form of saving a person requires the restriction, limitation, or even sacrifice of another person. This fundamental ethical issue will not be addressed in this chapter, since the present focus is on robogenesis, a technology that is (at least hypothetically) deliberately designed to maximize the viability and success of human reproduction.

[30] Aldous Huxley's brother, Julian, is usually credited with coining the term "transhumanism" (Bostrom 2005a, 6). Julian Huxley defined transhumanism as a belief that called for the transcendence of man and of the human species (Huxley [1957] 2015, 15). Christian Byk points out, however, that the term "transhumanism" had been used in the same context much earlier by the French philosopher Jean Coutrout (Byk 2021, 143).

of a new posthuman species (Bostrom 2005b, 207). The dispute within transhumanism then focuses on a disagreement about the definitive purpose of human enhancement. The more radical branch of transhumanism assumes that the ultimate goal of human enhancement is the transcendence of humans (Fernández and Rueda 2021, 226). Radical transhumanists therefore consider those forms of human modification that lead to the quickest elimination of the "human problem" (Mossbridge 2019, 302). The proponents of this type of transhumanism demand a fundamental acceleration of human transformation, which is why they argue in favour of the free and unrestricted use of any human enhancement (More [1999] 2013, 449). However, representatives of the more moderate branch of transhumanism argue that the qualitative transcendence of man is only a possible and thus not a necessary goal of human enhancement (Göcke 2018, 33). These more moderate transhumanists therefore argue in favour of the social regulation of human enhancement (Hughes 2004, 22). Proponents of moderate transhumanism thus assume that the most appropriate way to solve "the human problem" and create posthumans is through a gradual reform of human biology, which includes reforming human reproduction. From the point of view of transhumanism, robogenesis would be only one of the options that would lead (either rapidly or gradually) to the welcome elimination of fundamental human reproductive deficiencies. For transhumanists, the possible existence of robogenesis would also legitimize their philosophical approach to human beings; it would be evidence of the desirability of changing humans' basic biological characteristics. Both arguments in favour of applying robogenesis might thus be acceptable to transhumanists. Robogenesis would represent a technology that fundamentally increases the necessary space for the transformation of human biology and that constitutes another possibility in a series of steps (options) that may lead to the creation of posthumans. The possible compulsory application of robogenesis could in turn be seen by transhumanists as evidence of the necessary change of the human species into a posthuman one. Transhumanism would thus require a reformulation of the obligatory argument for the application of robogenesis. In a transhumanist lens, the effort to preserve the human species (humanism) would be replaced by the need to transcend the human species (posthuman species). The meaning of applying robogenesis is therefore different in transhumanism than it is in humanism. For transhumanism, the meaning of applying robogenesis would consist of proving the fragility of human beings, i.e. in proving an inappropriate ontology which precisely for this reason would need to be technologically transformed, improved, or abandoned. But even mastery of a new work does not necessarily mean that this work will ultimately have true value.

Posthumanism is an umbrella term for a wide range of philosophical theories that assume a human's value lies neither in his special past nor in a vision of his special future. This is simply because a human's value is not really special at all. Posthumanism is a group of critical theories that are based on the notion that the multitude of human shortcomings is definitive proof of humans' ordinariness. Posthumanists typically argue that this unremarkable character of humanity can be found in two basic ways of misunderstanding the true nature of humans. Firstly, humanism's visions of humankind's titanic past and present are simply false, because they do not coincide at all with humanity's actual agency (Ferrando 2019, 24). Posthumanists argue that the long-term agency of humanity is causing unprecedented environmental destruction, which can also lead to a threat to life itself. And the problem is not only the impact on the lives of other species (Valera 2014, 488). The same human approach is being applied within the human species itself. Many posthumanists point out that there are countless examples in human history where humans have been displaced from the community of humanity (Ferrando 2013, 28). For posthumanism, humanism is merely a misguided ideology that excuses human failings with fantasies of a "masterpiece". Such an illusion is hardly sustainable in the face of contemporary reality, which is why, according to posthumanists, there is a second kind of delusion about the nature of humanity today. The second way of obfuscating human nature is built on the projection of the idea of human exceptionalism into the future of posthumanity. Posthumanists argue that transhumanist visions of a Promethean future for posthumanity are as false as the assumptions of old humanism (Bolter 2016, 2). Transhumanists colonize the future instead of the past (Pearson 1997, 236). The dominant transhumanist orientation toward the future of posthumanity then leads to two fundamental flaws of transhumanism. The projection of humanity into the future of posthumanity leads to an ignoring of the demands of some groups of contemporary humanity and thus brings about an overlooking of the needs of people who are not fortunate enough to participate in the technical solutions to ensure the arrival of a new posthuman civilization (e.g. the problem of resource distribution). On the other hand, projecting humanity into a posthuman future also implies ignoring the demands of potential posthumans. Posthumanists point out that many transhumanist visions of the posthuman are built on the idea of maximizing current human characteristics and thus on the pursuit of human preservation in the posthuman (Roden 2010, 28). In doing so, transhumanism paradoxically limits the possibilities for the existence of different forms of posthumans. For posthumanism, transhumanism is therefore merely superhumanism, a concept that reiterates all the problems and failures of humanism in a superlative way. Posthumanists see the root of these common problems of humanism and transhumanism in the inappropriate exclusion of humanity from its relations to the world (Haraway 2003). In other words, the problem of humanism and transhumanism lies in the misconception of human identity. Posthumanists assume that human identity is not the result of some exceptional essence of humanity (humanism or transhumanism), but rather that it results from a complex intersection of diverse relationships between different aspects of the world (Haraway 1990, 197). A human is therefore exactly the same material object as all other objects of the world (e.g. new materialism). A human's identity is not closed, limited, or fixed, but is instead radically hybrid and open to all possible changes (Pisarski 2021, 3). Precisely because of this, it is also necessarily open to technological completion. The posthumanist attachment to technology is built on the idea of the dissolution of all forms of humankind's apparent exceptionalism; for posthumanism, there is no fundamental difference in principle between technologically enhanced and non-enhanced humans. In posthumanism, technology is only seen as a tool that increasingly demonstrates the untenable assumptions of humanism and transhumanism. Any forms of identity or assemblage of humans and technology are therefore permissible (Fox and Alldred 2020, 122). This even applies to those modes of assemblage that do not make any use of the technological completion of their identity. However, while all modes of being are equal, they are not the same (Braidotti 2020, 469). This also applies to diverse forms of reproduction. For posthumanism, robogenesis would be just one more possibility that points to the blurred boundaries between technology, humans, and other organisms (out-of-body reproduction); the voluntary use of robogenesis could serve as a fundamental example of the blurred boundaries of human biology and thus as a fundamental example of the merging of humans with posthumans. On the other hand, the obligatory form of robogenesis in posthumanism loses its justification, since the goal of posthumanism is not the necessary and exclusive preservation of humans or posthumans.[31] Otherwise, posthumanism would fall into similar problems to those it

[31] Posthumanism argues against the preference for the needs of some species (human animals) at the expense of the needs of others (non-human animals) (Schussler 2020, 40); however, equating the requirements of all species can paradoxically (under certain circumstances) also lead to ignoring the specific existential requirements of some species. For example, it is questionable whether posthumanism can formulate a criterion that would prevent the extinction of a particular species that threatens the survival of other species by its very existence (Bakošová and Odorčák 2020).

criticizes in humanism and transhumanism (i.e. a preference for only certain objects of the world). From the perspective of posthumanism, the point of applying robogenesis may be to illustrate the open ontology of the human, which is precisely why it is free to be supplemented by any technology.

Conclusion

It is quite obvious that the application of robogenesis for human reproduction would be associated with a whole series of very serious ethical, social, legal, and practical issues. For instance, it is not at all clear how the various procedures of robogenesis (e.g. automation and AI) would affect the complex debate on the ethical aspects of selection, cryopreservation, and embryo modification. Robogenesis would also likely have an impact on the related debate about the moral permissibility of certain biomedical interventions into the integrity of the human individual (automation of medicine). It would also fundamentally change certain social and gender expectations that are commonly associated with human reproduction (pregnancy); however, it could also increase the social inequalities that would result from the economic disparities between the benefits of robogenesis and the possibilities of normal reproduction (the problem of accessibility). From a legal point of view, the problem of robogenesis may in turn lie in the question of responsibility for carrying out processes that would completely automate and externalize human reproduction altogether. Ultimately, it is not at all clear even what date of birth would be entered on the birth certificate of an individual created by robogenesis. All of these serious (or curious) issues are mainly related to problems of the practical and technical safety of applying robogenesis. Since research is already underway into technologies related to some parts of the automation of human reproduction, it is reasonable to assume that the relevance of these fundamental practical issues will only increase.

On the other hand, the evaluation of robogenesis depends not only on the important practical implications of the application of this technology but also on substantive theoretical assumptions about the origins and value of the subject of robogenesis. From this perspective, the fundamental objection to robogenesis is the assumption that this technology could alter individuals' self-understanding, transform relationships between individuals, and modify continuity between generations. Such an argument would simply assert that robogenesis would be an artificial mode of reproduction, which would therefore create artificial humans. This naturalistic argument is based on a deeper ontological and epistemological problem that concerns the categorical distinction between artificial and natural objects;

however, the first problem with any naturalistic argument is that, for the most part, there is no clear criterion that definitively and indiscriminately just divides all objects and then only into artificial and natural ones. This does not naturally imply that natural and artificial objects do not exist, since in this case an absence of precise evidence is not evidence of absence. The more serious problem with the naturalistic argument is usually in the tacit premise that the artificial (whatever it may be) is simply bad. The reason for such an assessment is perhaps the idea that the artificial is somehow connected to humans, who are in many ways imperfect; therefore, anything artificial will eventually be imperfect and in many ways bad. From a certain point of view, such scepticism about the nature of humans is understandable. In the case of robogenesis, however, this argument is not entirely relevant, since it is based on the dilemma of choice between culture and nature. In the case of robogenesis, a third factor plays a crucial role: a culture which organizes itself naturally without human intervention (AI and automation). The specificity of robogenesis therefore lies in the fundamental personification of the trilemma of choice between nature, culture, and "natureculture".[32] Each trilemma provides more choices than each dilemma, and many different arguments for and against applying robogenesis are thus conceivable. This chapter has sought to show that the application of robogenesis could, under certain circumstances, co-create an ontological status for humans that is both more artificial (in the sense of technical intervention) and natural (in the sense of no human intervention). Such a paradoxical ontological consequence is also acceptable, to varying degrees, for contemporary theories of humanism, transhumanism, and posthumanism.

Acknowledgement
This chapter was written at the Department of Philosophy and Applied Philosophy, Faculty of Arts, University of Ss. Cyril and Methodius in Trnava, as a part of VEGA project 1/0563/18.

Bibliography

Afnan, Michael A. M., Cynthia Rudin, Vincent Conitzer, Julian Savulescu, Abhishek Mishra, Yanhe Liu, and Masoud Afnan. 2021. "Ethical Implementation of Artificial Intelligence to Select Embryos in Vitro Fertilization." Conference on AI, Ethics, and Society. *arXiv preprint*, arXiv:2105.00060. DOI: https://doi.org/10.1145/3461702.3462589.

[32] The concept of "natureculture" was introduced by Donna Haraway (2003, 11). Haraway defines natureculture as a synthesis of nature and culture that overcomes the dualism between them. In this chapter, however, natureculture refers to another category that can be added to nature and culture (trialism).

Annas, George J. 1984. “Making Babies Without Sex: The Law and the Profits.” *American Journal of Public Health* 74, 12: 1415–1417. DOI: https://doi.org/10.2105/AJPH.74.12.1415.

Anonymous. 1924. “(1) Daedalus, or Science and the Future (2) Icarus, or the Future of Science.” *Nature* 113: 740–741. DOI: https://doi.org/10.1038/113740a0.

Aristarkhova, Irina. 2005. “Ectogenesis and Mother as Machine.” *Body & Society* 11, 3: 43–59. DOI: https://doi.org/10.1177/1357034X05056190.

Bakošová, Pavlína, and Juraj Odorčák. 2020. “Posthumanism and Human Extinction: Apocalypse, Species, and Two Posthuman Ecologies.” *Journal for the Study of Religions and Ideologies* 19, 57: 47–62.

Ball, Philip. 2011. *Unnatural: The Heretical Idea of Making People*. London: Bodley Head.

Ball, Philip. 2013. “In Retrospect: Brave New World.” *Nature* 503: 338–339. DOI: https://doi.org/10.1038/503338a.

Benagiano, Giuseppe, and Maurizio Mori. 2009. “The Origins of Human Sexuality: Procreation or Recreation?” *Reproductive Biomedicine* 18: 50–59. DOI: https://doi.org/10.1016/S1472-6483(10)60116-2.

Berenbaum, May R. 2012. “Postlude.” *Daedalus* 141, 3: 121–124. DOI: https://doi.org/10.1162/DAED_a_00167.

Bernal, John Desmond. [1929] 2017. *The World, the Flesh and the Devil: An Enquiry into the Future of the Three Enemies of the Rational Soul*. EPUB, London: Verso.

Bolter, Jay David. 2016. “Posthumanism.” In *The International Encyclopedia of Communication Theory and Philosophy*, ed. by Jensen K. Bruhn and Robert T. Craig. EPUB, Chichester: Wiley & Sons.

Bostrom, Nick. 2005a. “A History of Transhumanist Thought.” *Journal of Evolution and Technology* 14, 1: 1–14.

Bostrom, Nick. 2005b. “In Defense of Posthuman Dignity.” *Bioethics* 19, 3: 202–214. DOI: https://doi.org/10.1111/j.1467-8519.2005.00437.x.

Braidotti, Rosi. 2020. “‘We’ Are in This Together, but We Are Not One and the Same.” *Bioethical Inquiry* 17: 465–469. DOI: https://doi.org/10.1007/s11673-020-10017-8.

Brittain, Vera. 1929. *Halcyon, or the Future of Monogamy.* London: Kegan Paul, Trench, Trübner & Co.

Bulletti, Carlo, and Carlos Simon. 2019. “Bioengineered Uterus: A Path Toward Ectogenesis.” *Fertility and Sterility* 112, 3: 446–447. DOI: https://doi.org/10.1016/j.fertnstert.2019.06.023.

Byk, Christian. 2021. “Transhumanism: From Julian Huxley to UNESCO: What Objective for International Action?” *Jahr – European Journal of Bioethics* 12, 1: 139–160.

Carbonell, Pablo, Tijana Radivojevic, and Héctor García Martín. 2019. "Opportunities at the Intersection of Synthetic Biology, Machine Learning, and Automation." *ACS Synthetic Biology* 8, 7: 1474–1477. DOI: https://doi.org/10.1021/acssynbio.8b00540.

Casciani, Valentina, Daniela Galliano, Jason M. Franasiak, Giulia Mariani, and Marcos Meseguer. 2021. "Are We Approaching Automated Assisted Reproductive Technology? Sperm Analysis, Oocyte Manipulation, and Insemination". *F&S Reviews* 2, 3: 189–203. DOI: https://doi.org/10.1016/j.xfnr.2021.03.002.

Chatzinikolaou, Nikolaos. 2010. "The Ethics of Assisted Reproduction." *Journal of Reproductive Immunology* 85, 1: 3–8. DOI: https://doi.org/10.1016/j.jri.2010.02.001.

Clark, Andy. 2013. "Re-Inventing Ourselves: The Plasticity of Embodiment, Sensing, and Mind." In *The Transhumanist Reader: Classical and Contemporary Essays on the Science, Technology, and Philosophy of the Human Future*, ed. by Max More and Natasha Vita-More, 113–127. Chichester: Wiley-Blackwell.

Cordeiro, José Luis. 2019. "The Boundaries of the Human: From Humanism to Transhumanism." In *The Transhumanism Handbook,* ed. by Newton Lee, 63–74. Cham: Springer.

Čapek, Karel. 1920. *R. U. R. (Rossum's Universal Robots).* Prague: Otokar Štorch-Marien Aventinum.

Daar, Judith. 2017. *The New Eugenics*. New Haven: Yale University Press.

De Gruchy, John W. 2018. "Christian Humanism, Progressive Christianity, and Social Transformation." *Journal for the Study of Religion* 31, 1: 54–69. http://dx.doi.org/10.17159/2413-3027/2018/v31n1a3.

Deech, Ruth, and Anna Smajdor. 2007. *From IVF to Immortality: Controversy in the Era of Reproductive Technology.* Oxford: Oxford University Press.

Derbyshire, Stuart. 2019. "The Biobag: A Brave New World, Part 1." *Conscience* 40, 2.

Earp, Brian D., and Julian Savulescu. 2020. *Love is the Drug: The Chemical Future of Our Relationships.* Redwood City: Stanford University Press.

Edwards, Matthew R. 2021. "Space Ectogenesis: Securing Survival of Humans and Earth Life with Minimal Risks – Reply to Szocik." *International Journal of Astrobiology* 20, 4: 323–326. DOI: https://doi.org/10.1017/S147355042100015X.

Ellis, Brian David. 2012. *Social Humanism: A New Metaphysics.* London: Routledge.

Felderhof, Marius. 2011. "Secular Humanism." In *Debates in Religious Education*, ed. L. Philip Barnes, 160–170. London: Routledge.

Ferber, Sarah, Nicola J. Marks, and Vera Mackie. 2020. *IVF and Assisted Reproduction: A Global History.* Singapore: Palgrave Macmillan.

Fernández, Belén Liedo, and Jon Rueda. 2021. "In Defense of Posthuman Vulnerability." *Scientia et Fides* 9, 1: 215–239. DOI: https://doi.org/10.12775/setf.2021.008.

Ferrando, Francesca. 2013. “Posthumanism, Transhumanism, Antihumanism, Metahumanism, and New Materialisms.” *Existenz* 8, 2: 26–32.

Ferrando, Francesca. 2019. *Philosophical Posthumanism*. London: Bloomsbury Publishing.

Ferreira, Aline. 2011. “Mechanized Humanity: JBS Haldane, JD Bernal, and Their Circle.” In *Discourses and Narrations in the Biosciences*, ed. by Brian Hurwitz and Paola Spinozzi, 119–130. Goettingen: V&R Unipress.

Ferreira, Aline. 2017. “The Fantasy of Ectogenesis in Interwar Britain: Texts and Contexts.” In *Exchanges Between Literature and Science from the 1800s to the 2000s: Converging Realms*, ed. by Márcia Lemos and Miguel Ramalhete Gomes, 136–154. Cambridge: Cambridge Scholars Publishing.

Figdor, Carrie. 2021. “The Psychological Speciesism of Humanism.” *Philosophical Studies* 178, 5: 1545–1569. DOI: https://doi.org/10.1007/s11098-020-01495-y.

Fox, Nick J., and Pam Alldred. 2020. “Sustainability, Feminist Posthumanism and the Unusual Capacities of (Post)Humans.” *Environmental Sociology* 6, 2: 121–131. DOI: https://doi.org/10.1080/23251042.2019.1704480.

Franklin, Sarah. 2013. *Biological Relatives: IVF, Stem Cells and the Future of Kinship.* Durham and London: Duke University Press.

Frith, Lucy. 2012. “Reproductive Technologies, Overview.” In *Encyclopedia of Applied Ethics (Second Edition)*, ed. by Ruth Chadwick, 766–774. London: Academic Press.

Garabedian, Michael. 1996. “A Humanistic Hamlet: Shakespeare’s Dissolution of the Courtly Prince.” *Literary Review* 10: 67–81.

Gehlen, Arnold. [1957] 1980. *Man in the Age of Technology*. New York: Columbia University Press.

Goyal, Ashish, Maheshwar Kuchana, and Kameswari P. R. Ayyagari. 2020. “Machine Learning Predicts Live-birth Occurrence Before In-vitro Fertilization Treatment.” *Scientific Reports* 10: 20925. DOI: https://doi.org/10.1038/s41598-020-76928-z.

Göcke, Benedikt. 2018. “Moderate Transhumanism and Compassion.” *Journal of Posthuman Studies* 2, 1: 28–44. DOI: https://doi.org/10.5325/jpoststud.2.1.0028.

Graur, Florin, Mihaela Frunza, Radu Elisei, Luminita Furcea, Liviu Scurtu, Corina Radu, Aron Szilaghy, Horatiu Neagos, Adriana Muresan, and Liviu Vlad. 2010. “Ethics in Robotic Surgery and Telemedicine.” In *New Trends in Mechanism Science. Mechanisms and Machine Science*, 5, ed. by Doina Pisla, Marco Ceccarelli, Manfred Husty, and Burkhard Corves, 457–465. Dodrecht: Springer.

Gromyko, Vladimir I., Valentina P. Kazaryan, Nicolay S. Vasilyev, Alexander G. Simakin, and *Stanislav* S. Anosov. 2018. “Artificial Intelligence as Tutoring Partner for Human Intellect.” In *Advances in Artificial Systems for Medicine and Education. AIMEE 2017. Advances in*

Intelligent Systems and Computing, 658, ed. by Zhengbing Hu, Sergey Petoukhov, and Matthew He, 238–247. Cham: Springer.

Gupta, Sahil. 2020. "AI Will Revolutionize Assisted Reproductive Technology (If We Work Together)." *Fertility and Sterility*, first published 15 July. Accessed 25 July, 2021. https://www.fertstertdialog.com/posts/ai-will-revolutionize-assisted-reproductive-technology-if-we-work-together.

Guzman, Andrea L. 2016. "Making AI Safe for Humans: A Conversation with Siri." In *Socialbots and their Friends: Digital Media and the Automation of Sociality*, ed. by Robert W. Gehl and Maria Bakardjieva, 69–85. New York: Routledge.

Hafer, Abby. 2020. "Humanism, Sex, and Sexuality." In *The Oxford Handbook of Humanism,* ed. by Anthony B. Pinn. EPUB, Oxford: Oxford University Press.

Haldane, John Burdon Sanderson. 1924. *Daedalus; or, Science and the Future*. London: Kegan Paul.

Haldane, John Burdon Sanderson. [1927] 2017. "The Last Judgment". In *Possible Worlds*, ed. by John Burdon Sanderson Haldane, 287–312 London: Routledge.

Haraway, Donna. 1990. "A Manifesto for Cyborgs: Science, Technology, and Socialist Feminism in the 1980s." In *Feminism/Postmodernism*, ed. by Linda Nicholson, 190–233. London: Routledge.

Haraway, Donna. 2003. *The Companion Species Manifesto: Dogs, People, and Significant Otherness. 1.* Chicago: Prickly Paradigm Press.

Haroon, Latif Khan. 2021. "New Insights into How AI Will Lead to Developments in Assisted Reproductive Technology." *News Medical,* 17 May. Accessed June 30, 2021. https://www.news-medical.net/news/20210517/New-insights-into-how-AI-will-lead-to-developments-in-assisted-reproductive-technology.aspx.

Hassan, Ihab. 1979. "Desire, Imagination, Change: Outline of A Critical Project." *Studies in the Literary Imagination* 12, 1: 129–143.

Helbing, Dirk. 2015. *The Automation of Society Is Next: How to Survive the Digital Revolution.* Scotts Valley: CreateSpace.

Hoc, Jean-Michel. 2007. "Human and Automation: A Matter of Cooperation." *HUMAN 07*: 277–285.

Holland, Ian, and Jamie A. Davies. 2020. "Automation in the Life Science Research Laboratory." *Frontiers in Bioengineering and Biotechnology* 8: 571777. DOI: https://doi.org/10.3389/fbioe.2020.571777.

Hotke, Andrew. 2014. "The Principle of Procreative Beneficence: Old Arguments and a New Challenge." *Bioethics* 28, 5: 255–262. DOI: https://doi.org/10.1111/j.1467-8519.2012.01999.x.

Huang, Kaicheng, Ihab Abu Ajamieh, Zhenxi Cui, Jiewen Lai, James K. Mills, and Henry K. Chu. 2021. "Automated Embryo Manipulation and Rotation via Robotic nDEP-Tweezers." *IEEE Transactions on Biomedical Engineering* 68, 7: 2152–2163.

Hughes, James. 2004. *Citizen Cyborg: Why Democratic Societies Must Respond to the Redesigned Human of the Future.* Cambridge: Westview Press.

Huxley, Aldous. [1921] 2018. "Crome Yellow." In *Complete Works of Aldous Huxley*. EPUB, Hastings: Delphi Classics.

Huxley, Aldous. [1932] 2018. "Brave New World." In *Complete Works of Aldous Huxley*. EPUB, Hastings: Delphi Classics.

Huxley, Aldous. 1963. "Utopias: Positive and Negative." *Proceedings of the American Academy of Arts and Letters and the National Institute of Arts and Letters, Second Series*, 13: 232–237.

Huxley, Julian. [1957] 2015. "Transhumanism." *Ethics in Progress* 6, 1: 12–16.

Jayakumaran, Jayapriya, Sejal D. Patel, Bhushan K. Gangrade, Deepa M. Narasimhulu, Soundarya R. Pandian, and Celso Silva. 2017. "Robotic-assisted Laparoscopy in Reproductive Surgery: A Contemporary Review." *Journal of Robotic Surgery*, 11: 97–109. DOI: https://doi.org/10.1007/s11701-017-0682-4.

Jeffreys, Mark. 2001. "Dr. Daedalus and His Minotaur: Mythic Warnings about Genetic Engineering from J. B. S. Haldane, François Jacob, and Andrew Niccol's Gattaca." *Journal of Medical Humanities* 22: 137–152. DOI: https://doi.org/10.1023/A:1009019712690.

Johnson, Jeffrey. 2008. "Science and Policy in Designing Complex Futures." *Futures* 40, 6: 520–536. DOI: https://doi.org/10.1016/j.futures.2007.11.012.

Khosravi, Pegah, Ehsan Kazemi, Qiansheng Zhan, Jonas E. Malmsten, Marco Toschi, Pantelis Zisimopoulos, Alexandros Sigaras, Stuart Lavery, Lee A. D. Cooper, Cristina Hickman, Marcos Meseguer, Zev Rosenwaks, Olivier Elemento, Nikica Zaninovic, and Iman Hajirasouliha. 2019. "Deep Learning Enables Robust Assessment and Selection of Human Blastocysts After In Vitro Fertilization." *npj Digital Medicine* 2: 21. DOI: https://doi.org/10.1038/s41746-019-0096-y.

Kigozi, Ronald. 2018. "Parental Responsibility and Assisted Reproductive Technologies." *Studia Bioethica* 11, 3: 41–49.

Kingma, Elselijn, and Suki Finn. 2020. "Neonatal Incubator or Artificial Womb? Distinguishing Ectogestation and Ectogenesis Using the Metaphysics of Pregnancy." *Bioethics* 34, 4: 354–363. DOI: https://doi.org/10.1111/bioe.12717.

Kopacek, Peter. 2019. "Trends in Production Automation." *IFAC-PapersOnLine* 52, 25: 509–512. DOI: https://doi.org/10.1016/j.ifacol.2019.12.595.

Kragh, Mikkel Fly, and Henrik Karstoft. 2021. "Embryo Selection with Artificial Intelligence: How to Evaluate and Compare Methods?" *Journal of Assisted Reproduction and Genetics* 38: 1675–1689. DOI: https://doi.org/10.1007/s10815-021-02254-6.

Kuru, Kaya, and Halil Yetgin. 2019. "Transformation to Advanced Mechatronics Systems Within New Industrial Revolution: A Novel Framework in Automation of Everything (AoE)." *IEEE Access*, 7: 41395–41415.

Lamm, Richard D. 1970. "The Reproductive Revolution." *American Bar Association Journal* 56, 1: 41–44.

LeMoncheck, Linda. 2020. "Philosophy, Gender Politics, and In Vitro Fertilization: A Feminist Ethics of Reproductive Healthcare." In *Women, Medicine, Ethics and the Law*, ed. by Susan Sherwin and Barbara Parish, 149–165. London: Routledge.

Lucke, Doris Mathilde. 2019. "Technologische Reproduktion." In *Die Maschine: Freund oder Feind?,* ed. by Caja Thimm and Thomas Christian Bächle, 333–380. Wiesbaden: Springer VS.

Luckhurst, Roger. 2014. "Automation." In *The Oxford Handbook of Science Fiction*, ed. by Rob Latham, 317–328. Oxford: Oxford University Press.

Ludovici, Anthony. 1924. *Lysistrata, or Women's Future and Future Women*. London: Kegan Paul, Trench, Trubner and Co.

MacInnes, John, and Peréz Díaz. 2009. "The Reproductive Revolution." *The Sociological Review* 57, 2: 262–284. DOI: https://doi.org/10.1111/j.1467-954X.2009.01829.x.

Marovitzs, Sanford E. 2003. "Aldous Huxley and the Nuclear Age: Ape and Essence in Context." In *Bloom's Modern Critical Views: Aldous Huxley*, ed. by Harold Bloom, 137–152. New York: Chelsea Publishing.

Meckier, Jerome. 2003. "Aldous Huxley, Satiric Sonneteer: The Defeat of Youth." In *Bloom's Modern Critical Views: Aldous Huxley*, ed. by Harold Bloom, 85–108. New York: Chelsea Publishing.

Melhi, Elnaz, Fatemeh Bagherian, Golnaz M. N. Fard, and Niloufar Farsijani. 2020. "From Existential-humanism Approach to Frankl's Logotherapy." *Rooyesh-e-Ravanshenasi Journal* 9, 4: 171–180.

Meseguer, Marcos, Ulrich Kruhne, and Steen Laursen. 2012. "Full In Vitro Fertilization Laboratory Mechanization: Toward Robotic Assisted Reproduction?" *Fertility and Sterility* 97, 6: 1277–1286. DOI: https://doi.org/10.1016/j.fertnstert.2012.03.013.

Miyagi, Yasunari, Toshihiro Habara, Rei Hirata, and Nobuyoshi Hayashi. 2019. "Feasibility of Predicting Live Birth by Combining Conventional Embryo Evaluation with Artificial Intelligence Applied to a Blastocyst Image in Patients Classified by Age." *Reproductive Medicine and Biology* 18, 4: 344–356. DOI: https://doi.org/10.1002/rmb2.12284.

Montgomery, Marion. 1974. "Lord Russell and Madame Sesostris." *The Georgia Review* 28, 2: 269–282.

Moran, Margaret. 1984. "Bertrand Russell as Scogan in Aldous Huxley's 'Crome Yellow'." *Mosaic: A Journal for the Interdisciplinary Study of Literature* 17, 3: 117–132.

More, Max. [1999] 2013. "Letter to Mother Nature." In *The Transhumanist Reader: Classical and Contemporary Essays on the Science, Technology, and Philosophy of the Human Future*, ed. by Max More and Natasha Vita-More, 449–450. Chichester: Wiley-Blackwell.

Mossbridge, Julia A. 2019. "Hacking the Human Problem." In *The Transhumanism Handbook*, ed. by Newton Lee, 301–310. Cham: Springer.

Nasiri, Nahid, and Poopak Eftekhari-Yazdi. 2015. "An Overview of the Available Methods for Morphological Scoring of Pre-implantation Embryos in In Vitro Fertilization." *Cell Journal* 16, 4: 392–405.

Neresini, Federico. 2011. "Social Aspects of Biobanking: Beyond the Public/Private Distinction and Inside the Relationship Between the Body and Identity." In *Biobanks and Tissue Research: The Public, the Patient and the Regulation*, ed. by Christian Lenk, Judit Sándor, and Bert Gordijn, 65–78. Dordrecht: Springer.

Norman, Richard. 2004. *On Humanism*. London: New York.

Nowak, Rachel. 2007. "A Reproductive Revolution." *New Scientist* 193, 2596: 8–9. DOI: https://doi.org/10.1016/S0262-4079(07)60707-6.

Nowottny, Winifred. 1964. "Shakespeare's Tragedies." In *Shakespeare's World*, ed. by James Sutherfield and Joel Hurtsfield, 48–78. New York: St. Martin's Press.

Odorčák, Juraj. 2020. "Robotické bábätká." In *100 let R.U.R.*, ed. by Peter Jemelka and Slavomír Lesňák, 47–60. Brno: Masaryk University Press.

Odorčák, Juraj, and Pavlína Bakošová. 2021. "Robots, Extinction, and Salvation: On Altruism in Human–Posthuman Interactions." *Religions* 12, 4: 275. https://doi.org/10.3390/rel12040275.

Ouyang, Fan, and Pengcheng Jiao. 2021. "Artificial Intelligence in Education: The Three Paradigms." *Computers and Education: Artificial Intelligence* 2, 100020: 1–6. DOI: https://doi.org/10.1016/j.caeai.2021.100020.

Pang, Zhibo, Geng Yang, Ridha Khedri, and Yuan-Ting Zhang. 2018. "Introduction to the Special Section: Convergence of Automation Technology, Biomedical Engineering, and Health Informatics Toward the Healthcare 4.0." *IEEE Reviews in Biomedical Engineering* 11: 249–259.

Partridge, Emily A. et al. 2017. "An Extra-uterine System to Physiologically Support the Extreme Premature Lamb." *Nature Communications* 8, 1: 1–16. DOI: https://doi.org/10.1038/ncomms15112.

Pearson, Keith Ansell. 1997. "Life Becoming Body: On the 'Meaning' of Post-human Evolution". *Cultural Values* 1, 2: 219–224. DOI: https://doi.org/10.1080/14797589709367145.

Pennings, Guido, and Guido De Wert. 2003. "Evolving Ethics in Medically Assisted Reproduction." *Human Reproduction Update* 9, 4: 397–404.

Pisarski, Mariusz. 2021. "Human, Super-human, Anti-human: The Posthuman Deep Future in Evolutionary Science Fiction." *World Literature Studies* 1, 13: 3–17. DOI: https://doi.org/10.31577/WLS.2021.13.1.1.

Prudil, Lukas, and Ladislav Pilka. 2002 "Legal and Ethical Problems in Assisted Reproduction: Case Report." *Česká gynekologie* 67, 3: 174–177.

Rähme, Boris. 2021. "Is Transhumanism a Religion?" In *Religion in the Age of Digitalization: From New Media to Spiritual Machines*, ed. by Giulia Isetti, Elisa Innerhofer, Harald Pechlaner, and Michael de Rachewiltz, 119–134. New York: Routledge.

Robey, Bryant, Shea O. Rutstein, and Leo Morris. 1992. "The Reproductive Revolution: New Survey Findings." *Population Reports* 20, 4: 1–44.

Roden, David. 2010. "Deconstruction and Excision in Philosophical Posthumanism." *The Journal of Evolution & Technology* 21, 1: 27–36.

Romanis, Elizabeth C., Dunja Begović, Margot R. Brazier, and Alexandra K. Mullock. 2020. "Reviewing the Womb." *Journal of Medical* Ethics, July 29, 2020. DOI: http://dx.doi.org/10.1136/medethics-2020-106160.

Ross, Fiona C., and Tessa Moll. 2020. "Assisted Reproduction: Politics, Ethics and Anthropological Futures." *Medical Anthropology* 39, 6: 553–562. DOI: https://doi.org/10.1080/01459740.2019.1695130.

Russell, Bertrand. 1924. *Icarus, or, the Future of Science*. London: Kegan Paul.

Russell, Bertrand. [1931] 2001. *The Scientific Outlook*. London: Routledge.

Russell, Bertrand. [1932] 1997. "Review in New Leader. 11 March 1932, p. 9." In *Aldous Huxley*, ed. by Donald Watt, 210–212. London: Routledge.

Saunders, Max. 2019. *Imagined Futures: Writing, Science, and Modernity in the To-Day and To-Morrow Book Series, 1923–31*. Oxford: Oxford University Press.

Savulescu, Julian. 2001. "Procreative Beneficence: Why We Should Select the Best Children." *Bioethics* 15, 5–6: 413–426. DOI: https://doi.org/10.1111/1467-8519.00251.

Schlick, Jochen. 2012. "Cyber-physical Systems in Factory Automation: Towards the 4th Industrial Revolution". In *2012 9th IEEE International Workshop on Factory Communication Systems*, ed. by Thomas Nolte and Andreas Willig, 55–65. New York: IEEE.

Scholl, Keller, and Robin Hanson. 2020. "Testing the Automation Revolution Hypothesis." *Economics Letters*, 193: 109287. DOI: https://doi.org/10.1016/j.econlet.2020.109287.

Schussler, Aura-Elena. 2020. "Posthumanism and Ecofeminist Theology: Toward a Nondualist Spirituality." *Journal for the Study of Religions and Ideologies* 19, 57: 32–46.

Segers, Seppe. 2021. "The Path Toward Ectogenesis: Looking Beyond the Technical Challenges." *BMC Medical Ethics* 22: 59. DOI: https://doi.org/10.1186/s12910-021-00630-6.

Setiya, Kieran. 2018. "Humanism." *Journal of the American Philosophical Association* 4, 4: 452–470. DOI: https://doi.org/10.1017/apa.2018.38.

Shakespeare, William. [1600–1601] 2000. "Hamlet." In *Cliffs Complete Shakespeare's Hamlet*, ed. by Sidney Lamb, 31–197. New York: Hungry Minds.

Singer, Peter, and Deane Wells. [1985] 2006. "Ectogenesis." In *Ectogenesis: Artificial Womb Technology and the Future of Human Reproduction*, ed. by Scott Gelfand and John R. Shook, 9–26. Amsterdam: Rodopi.

Smith, Grover, 1969. *Letters of Aldous Huxley*. London: Chatto & Windus.

Smith, Jim L. 2018. "The Effects of Automation." *Quality* 57, 7: 13.

Sroga, Julie, Sejal D. Patel, and Tommaso Falcone. 2008. "Robotics in Reproductive Medicine." *Frontiers in Bioscience* 13: 1308-1317. DOI: https://doi.org/10.2741/2763.

Thody, Philip. 1973. *Aldous Huxley*. London: Studio Vista.

Tomašovičová, Jana. 2021. "Parallels Between Two Worlds: Literary Science-Fiction Imagery and Transhumanist Visions." *World Literature Studies* 13, 1: 31–42. DOI: https://doi.org/10.31577/WLS.2021.13.1.3.

Tripp, Ronja. 2015. "Biopolitical Dystopia: Aldous Huxley, Brave New World (1932)." In *Dystopia, Science Fiction, Post-Apocalypse: Classics – New Tendencies – Model Interpretations*, ed. by Eckart Voigts and Alessandra Boller, 29–45. Trier: Wissenschaftlicher Verlag Trier.

Trolice, Mark P., Carol Curchoe, and Alexander M. Quaas. 2021. "Artificial Intelligence: The Future Is Now." *Journal of Assisted Reproduction and Genetics* 38: 1607–1612. DOI: https://doi.org/10.1007/s10815-021-02272-4.

Twine, France Winddance. 2015. *Outsourcing the Womb: Race, Class, and Gestational Surrogacy in a Global Market*. London: Routledge.

Valera, Luca. 2014. "Posthumanism: Beyond Humanism?" *Cuadernos de Bioética* 25, 3: 481–491.

Valerio, Carlos, Karen Vargas, and Henriette Raventós. 2017. "IVF in Costa Rica." *JBRA Assisted Reproduction* 21, 4: 366–369.

Vallverdú, Jordi, and Sarah Boix. 2019. "Ectogenesis as the Dilution of Sex or the End of Females?" In *Feminist Philosophy of Technology*, ed. by Janina Loh and Mark Coeckelbergh, 105–122. Berlin: J. B. Metzler.

Varghese, Alex C., and Charalampos S. Siristatidis. 2019. "Automation, Artificial Intelligence and Innovations in the Future of IVF." In *In Vitro Fertilization*, ed. by Zsolt Peter Nagy, Alex C. Varghese, and Ashok Agarwa, 847–860. Cham: Springer.

VerMilyea, Mathew, Jonathan M. M. Hall, Sonya M. Diakiw, Adrina Johnston, Tuc Van Nguyen, Donato Perugini, Andy Miller, Alicia Picou, Andrew Murphy, and Michelle Perugini. 2020. "Development of an Artificial Intelligence-based Assessment Model for Prediction of

Embryo Viability Using Static Images Captured by Optical Light Microscopy During IVF." *Human Reproduction* 35, 4: 770–784. DOI: https://doi.org/10.1093/humrep/deaa013.

Von Miese, Ludwig. 1944. *Bureaucracy*. New Haven: Yale University Press.

Wang, Renjie, Wei Pan, Lei Jin, Yuehan Li, Yudi Geng, Chun Gao, Gang Chen, Hui Wang, Ding Ma, and Shujie Liao. 2019. "Artificial Intelligence in Reproductive Medicine." *Reproduction* 158, 4: 139–154. DOI: https://doi.org/10.1530/REP-18-0523.

Wilkinson, Dominic, and Lydia Di Stefano. 2020. "Artificial Gestation." In *Emerging Topics and Controversies in Neonatology*, ed. by Elaine M. Boyle and Jonathan Cusack, 43–55. Cham: Springer.

Wilson, Daniel H. 2014. *Robogenesis*. New York: Doubleday.

Wilson, Duncan. 2011. *Tissue Culture in Science and Society: The Public Life of a Biological Technique in Twentieth-century Britain*. New York: Palgrave Macmillan.

Wirtz, Jochen, Werner Kunz, and Stefanie Paluch. 2021. "The Service Revolution, Intelligent Automation and Service Robots." *European Business Review* January/February: 38–44.

Zaninovic, Nikica, and Zev Rosenwaks. 2020. "Artificial Intelligence in Human In Vitro Fertilization and Embryology." *Fertility and Sterility* 114, 5: 914–920. DOI: https://doi.org/10.1016/j.fertnstert.2020.09.157.

Zhe, Lu, Xuping Zhang, Clement Leung, Navid Esfandiari, Robert F. Casper, and Yu Sun. 2011. "Robotic ICSI (Intracytoplasmic Sperm Injection)." *IEEE Transactions on Biomedical Engineering* 58, 7: 2102–2108. DOI: https://doi.org/10.1109/TBME.2011.2146781.

Zimmer, Katarina. 2021. "Artificial Wombs are Science Fiction: But Artificial Placentas are on the Horizon." *IEEE Spectrum* 58, 4: 22–29. https://doi.org/10.1109/MSPEC.2021.9393995.

Chapter 2

Is It Still Me? The Self, Memory, and the Relevance of the First-Person Perspective

Andrej Rozemberg

Abstract: What would really change if it turned out that a person is nothing more than a sequence of mental events and that the permanent self is an illusion? Of course, the answer is "nothing". Some contemporary authors have attempted to resolve this paradox, not dissimilarly to early Abhidharma scholars, by distinguishing between conventional and metaphysical levels of argumentation. On the conventional or phenomenal level, it would be absurd to deny the reality of persons or subjects of experience; however, outside of that – independently of the facts of our grammar or language – there are no such things as selves or persons. In this chapter, I argue that such a reductionist division is unjustified for many reasons. Despite the fact that non-self theories are unable to weaken the declared illusiveness of our sense of self, and parasitize often on our first-person intuitions and language (while simultaneously denying their ontological claims), there is no good reason to rule out the self from the debate of personal identity. Particularly if this self – being phenomenologically immune to psychological and bodily changes – seems to match our sense of diachronic identity and face several theoretical difficulties (e.g. "replacement", "fission", "duplication", or "memory gaps") better than reductionist approaches that analyse identity in terms of psychological continuity. I argue that although selves may not be the same thing as persons in an obvious sense, they should not be neglected as the primary source of our sense of identity over time. Using John Locke's memory theory of personal identity, I illustrate the theoretical difficulties that can arise from confusing the identity of the rememberer with the continuity of memories. I argue, *inter alia,* that if memory is what constitutes our identity, then there are necessarily persons without a past. On the contrary, if we accept the possibility of forgetful subjects, we can think under certain conditions about prudential concern and moral responsibility even in the absence of autobiographical memories, regardless of whether the narrative selves are separated by retrograde amnesia or physical death.
Keywords: Self, sense of self, first-person perspective, reductionism, memory, memory impairments, personal identity.

> Just as a calf finds its own mother among thousands of cows, so actions done in a former life unerringly reach the perpetrator thereof.
> *Viṣṇu Dharmasūtra XX*. 47

Introduction

In the *Devadūta Sutta* of the *Majjhima Nikāya*, Yama, the god of death, explains to a certain unfortunate soul who finds himself in a place of suffering (*nāraka*) that it was he who had committed a grave sin and was now condemned to taste its bitter fruits.[33] For orthodox Buddhists, this passage poses a considerable exegetic problem. The main reason for this lies in a theory according to which there is no self behind the changing mental and bodily states and that what is called a "person" for purely conventional reasons is only a bundle of subpersonal constituent components (*skandha*), or – in a diachronic sense – a series of momentary entities.[34] But when one returns to the opening sutra, one finds that the god of death does not think like a Buddhist philosopher. He does not say to that person that it was the "earlier stage of the causal chain" that had performed the act and that the "later stage" bore responsibility. Nor does he say that he is "neither the same nor a different person" (*na ca so, na ca añño*).[35] He says that "it was he" (*tayā v' etaṃ*), which means that this person had to have existed at the time of the given deed, even without remembering it.

In this regard, it is worthwhile pointing out a certain interesting moment in the broader historical and philosophical context, namely, that the Buddhist "non-self" doctrine, in addition to being probably the first non-intuitive theory of the person, is also an example of the powerlessness of philosophical theories in the face of lived reality. One remarkable example of this powerlessness can be seen in the story of the monk Khemaka (SN 22.89, S III,130), who even after years of study is unable to rid himself of his illusory sense of self. Khemaka responds to the questions of older monks of Kosambi by saying that although he does not consider any of the skandhas to be the self, he still cannot get rid of the concept of "I am" (*asmï ti*). A more interesting thing than this statement, however, is the solution (or trick) by which some authors have attempted to mitigate that paradox, and which, with a touch of irony, may be called the "perfect trick of reductionism". This solution consists of two steps. The first of these is to distinguish between two "levels" or truths: (i) the phenomenal or conventional and (ii) the metaphysical. The second step is to claim that the concept of the "non-self" does not deny the reality of persons at the phenomenal level; it merely does not recognize them as ontologically fundamental beings. Even if a reductionist philosopher knows that there is no self, in ordinary life he may nonetheless refer to himself in the first-person mode, use a name, take on commitments, make promises, and possess things, which is a strategy that does not cause Buddhist authors any serious problems, as Steven Collins (1994) writes in an article entitled "What are Buddhists Doing When They Deny the Self?" The Oxford philosopher Derek

[33] *tayā v' etaṃ pāpaṃ kammaṃ kataṃ; tvañ ñeva tassa vipākaṃ paṭisaṃvedissasīti*, MN 130, M iii 178.
[34] In the words of *Visudhimagga* XVI.90: "For there is suffering, but none who suffers; Doing exists although there is no doer..."
[35] See Minh Châu (1964, 61).

Parfit says it almost identically to the Buddhist philosophers, merely emphasizing the pragmatic aspect a little more:

> An outright denial [of a person – AR] is of course absurd. As Reid protested in the eighteenth century, "I am not thought, I am not action, I am not feeling. I am something which thinks and acts and feels." […] A Bundle Theorist admits this fact, but claims it to be only a fact about our grammar, or our language. There are persons or subjects in this language-dependent way. If, however, persons are believed to be more than this […] the Bundle Theorist denies that there are such things. (Parfit [1987] 2016, 93)

What does such a division mean from a methodological point of view? First of all, it means that in the given scheme of things – and this applies to all reductionist approaches – there will be a logical preference for approaches that analyse the identity of a person in terms of continuity (bodily, psychological, or phenomenal) and that reduce identity to something that it is not. Such an analysis of persons will often consist of descriptions of mental and physical states or events without referring to the subject of these states, who can be judged to be something more than the product of language or autobiographical memory; furthermore, this purely impersonal description of our lives will be taken as being complete (Parfit 1984, 341).

The position I argue for here will be quite different. It will be an approach that seriously considers the first-person sense of self, even though it will not, and cannot in principle, meet the required third-person criteria. It is this self – an irreducible subject of experience which persists through changes of qualities – that makes the question of personal identity a real problem. And it is this self that is, as I will try to show, a more plausible account of our sense of identity over time than the continuity of memories or R-relation.

In defence of the noncriterial approach, I would like to point out that the absence of epistemic criteria, especially non-circular ones, can sometimes result from the nature of things. Let us suppose my ordinary sense of cross-temporal identity includes periods separated from the present not by dozens but rather by thousands of gaps in consciousness. Nonetheless, I do not feel the need to ask: "How do I know I am the same person?" or "Upon what basis do I identify myself with the person of my memories?" To remember *x* does not mean to firstly judge whether I am the one who experienced *x* and only then conclude that I properly remember *x*. In other words, from a first-person perspective, I approach identity over time in a non-inferential and non-criterial manner. Moreover, but this does not concern the problem of circularity, it seems that narrowing identity criteria to identification criteria can be risky. We can imagine situations in which we successfully identify person *S* using physical or even memory criteria, but we can be mistaken about what constitutes his identity. A well-known example is Locke's memory theory of personal identity, which dismisses the thinking substance from the identity game on the grounds that it does not meet identification criteria. One section deals with this matter, wherein I attempt to show how several of its conclusions are possible only because of a conflation of the epistemological and ontological level of argumentation, and how this conflation can lead to a dead end. I will also try to justify why the diachronic self – which I interpret as the substantive self[36] – is a plausible solution to the "bridge problem" and a suitable approach to the problem of personal identity over time. In addition to its intuitiveness and its ability to plausibly account for phenomena such as recollection, desire, and anticipation or concern for the future, one advantage of such an approach is its ability to counter the theoretical difficulties associated with things such as "replacement", "fission", "duplication", "memory gaps", or "interruptions of stream of consciousness", which is where reductionist theories commonly fail. In the last section, I return to some practical implications of psychological theories of personal identity, specifically the thesis of the absence of responsibility and rational concern for the future in the absence of memories. I argue that since memory does not constitute a person's identity, absent memories may be compatible with the idea of justice and practical concerns under certain conditions.

The relevance of the first-person perspective

There are several objective reasons why the question of personal identity over time (hereafter PI) has persistently resisted attempts to resolve it. The first is the confusion of different levels of argumentation. If I was to utter the sentence "That's the man I saw in the theatre yesterday," I do not have to be a proponent of a bodily theory of identity, nor do I have to doubt the relevance of other ontological "criteria" of identity in order to identify the person reliably. Naturally, the problem does not usually arise in the ordinary recognition of other people; in the world that we live in, persons do not freely exchange bodies, create indistinguishable replicas, or branch into an infinite number of psychological continuants as they do in the thought experiments of philosophers. On the contrary, the kind of

[36] Despite some scepticism about the notion of the self as a substance, there are several well-known reasons for such an approach, from the monadic character of this self; the fact that frequent gaps in the stream of consciousness, such as a dreamless sleep, cannot weaken our awareness of this enduring self; through to the question of ownership of thoughts, experiences, and so on. Since thoughts are not thinking and consciousness alone is not conscious, it is not unjustified to infer an underlying subject as an experiencer, thinker of thoughts, and agent of actions, that is, the self as a bearer of properties that possesses and phenomenally unites its experiences.

beings that we call "persons" are characterized by a remarkable constancy in both a bodily and a psychological sense; persons do not undergo radical changes in body or character, at least not routinely nor suddenly. This is probably the main reason why we rely almost exclusively on third-person criteria, even when we do not identify persons with bodies or series of mental states. In fact, we do not have any other choice given that we can only have experience of the other self from the "outside". If my face is captured on some security camera footage, I could hardly succeed in any appeal by referring to the incorrectness of the bodily criterion of PI. However, this does not mean that "the same body" or "bodily spatio-temporal continuity", if a scenario such as that described by Mark Twain in *The Prince and the Pauper* were to occur, would be the correct answer to the question of what constitutes PI. In other words, it could be risky to reduce the problem of PI to the epistemological one. Let us assume for a while the opposite statement. One could argue, as Jay Rosenberg once did, that "[o]ur ability to apply any notion of personal identity at all is parasitic upon the existence of a conceptual apparatus used for individuating, identifying, and re-identifying of objects, causally interacting substances in space as well as time" (1981, 151). In other words, for it to make sense to consider the identity of *x* across time, *x* must be a publicly accessible spatio-temporal object. Otherwise, it could not be identifiable and the notion of identity would be rendered meaningless. Such an approach would, of course, prohibit any Lockean body-swaps and indeed all events and entities that do not satisfy the observability criterion. If such events and entities were metaphysically possible, proponents of the third-person approach could easily be mistaken in their recognitions. More precisely, it would not prohibit such entities but only their identification with persons.

The question is what follows from this narrowing of the problem. Let us suppose, contrary to Rosenberg's thesis of irrelevance of the first-person perspective for PI, that we wanted to insist that the first-person account is still relevant for PI because of the sense of identity which we normally have as persisting subjects of experience. If one were to object that in such a case it makes little sense to think about personal *identity*, given the absence of third-person epistemic criteria, we could argue that our notion of PI is non-criterial. If, on the other hand, one were to point out that we are using the notion of the *person* incorrectly, we could argue that the ordinary concept of the person derived from everyday social practice is not comprehensive enough to integrate our sense of diachronic identity as something, which is phenomenologically immune to changes of persons as spatio-temporal objects.

However, there is another important reason why it might be wise to consider the first-person perspective. Let us assume that those who think that the primary motive behind our interest in PI is the question of survival are right. It may even be the kind of survival that John Searle doubts when he asks whether it is necessary to "postulate the existence of a self that goes beyond the recognition of the body and of the sequence of experiences that occur in the body" (2005, 7). Admittedly, I do not wish to diminish the importance of some ethical issues related to the problem of identity, such as brain transplantation, DBS, the reconsolidation of memories, and the moral responsibility of amnesiacs. I just think (and I agree with Parfit on this) that the primary motive determining the vector of thinking about identity is the question of survival: Can I survive the permanent loss of memories? Or a brain transplant? Or physical death? Consider the following situations: (i) we are told by our doctor that we are going to have a difficult operation – we will "survive", but we will not remember any episode from our past life; and (ii) we will "not survive", but with the help of advanced technology our psychological profile (including our memories) will be copied into another body (brain). Which of these situations describes what we mean by our survival? The answer to this question will understandably depend on what one means when talking about the self and survival. It is here that the difference between the two perspectives becomes apparent, because approaches that reduce identity to continuity will not be interested in the nature of this self. They will not ask what x is but rather how x can be identified as the same entity across time. When Parfit and Shoemaker ask the question "What must apply in order to say that x at time t and y at time $t1$ are numerically the same person?" they are asking about the criteria for (re)identification. Here, to survive is to pass the identification test. The problem, however, is that the self – being unanalysable in terms of bodily or psychological continuity – provides no informative criteria of persistence. From a first-person perspective, I do not even need them. I am non-inferentially aware of this enduring self, and this awareness, as simple-view philosophers would say, seems to be independent of any knowledge of properties.[37] I do not identify myself as I identify others, say, upon the basis of bodily criteria. It is not the case that I have to look first in the mirror to make sure that it is me, or that I would observe some subject-

[37] In his *Two Selves*, Stanley Klein puts it very similarly to as Geoffrey Madell and Richard Swinburne when he writes that when he uses the expression "sense" of self to describe "our experiential acquaintance with the ontological self (i.e., the subject of experience)", he is "trying to convey a form, or aspect, of experience that is pre-reflectively felt 'as myself'; that is, an experience taken *directly* without the need for inference or the need to refer to, acknowledge, or recognize the content of the experience. It cannot be thematized or otherwise analysed; we are acquainted with it directly as a content-free feeling" (2014, 14).

neutral psychological characteristics upon the basis of which I could then identify as me. Circularity here is inescapable. Geoffrey Madell put it very similarly when he says that although "I am aware of myself as having certain properties [...] I do not and cannot identify myself through observing certain properties whose character indicates that they are mine." (2015, 5) What is here of special importance, however, is that although I have no direct phenomenological experience of a cross-temporal identity of the self as something that lasts "from thought to thought" (Strawson 2017, 35), frequent gaps in the stream of consciousness, such as dreamless sleep, cannot weaken my awareness of this enduring self just as changes within our minds and bodies are incapable of so doing. Squire et al. (1981), Tulving (1993), Klein et al. (2002), Rathbone et al. (2014), and Dorahy et al. (2021) refer to cases of patients suffering from severe memory impairments, dissociative identity disorder, and cognitive impairments, who despite a loss of "access to a variety of self-relevant sources of knowledge" (Klein 2012, 478) possessed a coherent sense of self which had not collapsed under the weight of cognitive disorders. But if this is the case, then terms like the "loss of self",[38] "damaged self", or "loss of sense of personal identity" are rather hyperbole and constructions suffering from a lack of distinction between the "self" and "self-concept", "self-image", "the autobiographical self", "the narrative self", and so on. In *The Self and its Brain* (2012, 474) Stanley Klein calls this permanent self the "ontological self" (the self of first-person subjectivity) in order to distinguish it from the self-object – the person with emotions, an individual life history, and social relationships. For sure, one could propose a different "division", or avoid any duplication of the self, but this does not change the main argument. What is essential is that no other known fact concerning our bodies and minds corresponds more adequately to our notion of identity than this permanent sense of the self. I

[38] Probably the most notable in this connection are cases of DPD, which are characterized by a feeling of loss of personal ownership or any attribution of bodily or mental states to the self. Daphne Simeon and Jeffrey Abugel (2006) give us many examples of such states: "At times his arms and legs feel like they don't belong with his body [...] His mind feels like it is operating apart from his body" (2006, 5); "It's like my thoughts are on a big movie screen" (9); "[B]ut I just disappeared inside. I went to a state of nothingness, no mood at all, as if I were dead" (30). Nonetheless, the question remains of whether we are right in interpreting such states as a "loss of self". It is not clear from the given description that what DPD patients experience is a missing subject, although we can interpret them as a missing sense of personal ownership of one's mental or physical states. After all, several uses of the pronoun "I" by the quoted patients, just like the fact that they perceive their situation as a misfortune and wish to change it (which would be understandable if there was nobody who perceives, wishes, and so on), indicates that what is missing here is not the experiential subject itself. As Stanley Klein interprets it, "[i]n such cases it appears that intact self-referential content exists in conjunction with functioning first-person subjectivity, albeit a subjectivity bewildered by the absence of felt ownership of the content of its experiences" (2015, 365).

therefore suggest that there is no reason why the first-person perspective should be overlooked in the search for an answer to the question of whether I will survive.

Naturally, I am aware that there has been considerable scepticism about first-person argumentation in philosophical debates on PI for quite some time. Nonetheless, the epistemic uncertainty that seems to be its enduring feature and reason of this scepticism is not the only problem being faced here. Another problem A – which phenomenon that we could eliminate without much intellectual effort and which is a pragmatic feature of argumentation – is the confusion of perspectives. The phenomenon of the confusion of perspectives is significant for the ability to create a semblance of clarity and apparent plausibility even in realistically impossible situations. If one changes perspectives and respects their specific rules (including linguistic ones) – i.e. instead of asking "What happens when I divide?" (Parfit 1984, 253), one would use, say, the phrase "What will happen if I undergo a hemispherectomy?" or "If half of my brain is removed from my body and surgically fused with the body of person *Y*..." It is thus possible that the description would be more accurate yet less persuasive since there is no picture of the divided self from the second version of the description. To get such a picture, we would have to assume a number of non-self-evident things, such as the identity of the self and the brain, or that with the self, which is monadic in nature, we can do what we do with bodies or body parts. Here is a brief example: some time ago, a parent of a hemispherectomy patient approached David Chalmers with the question, or rather concern, of whether a "second consciousness" or even a second person might have formed in his son's split brain, which could thus affect the quality of his life. And if this was indeed the case, would they have a moral obligation to that person? According to one possible interpretation (Schechter 2015), the severing of the corpus callosum gives rise to two non-communicating streams of consciousness, or subpersonal conscious systems, with their own experiences and cognitions, which are unaware of each other; however, both of them may retain memories of the original person, who they continue to identify with. For some philosophers, this dual interpretation presented an opportunity to definitively challenge the unitary subject of experience, or at least the phenomenal unity of consciousness (see e.g. Nagel 1971 and Parfit 1984). If there are two independent streams of consciousness, then there is no self or subject of experience that clearly rules over all mental states, which means, as Parfit believes, that our personal existence is not a matter of all or nothing. In hindsight, however, it appears that things may be different, since interpretations of agnosia and apraxia in hemispherectomy patients known from the earliest studies (Gazzaniga et al.

1967; Sperry 1968), which Parfit drew upon on in *Reasons and Persons*, have been revised several times, even by the authors themselves (Gazzaniga 1989; Savazzi and Marzi 2004; Pinto et al. 2017; de Haan et al. 2020). According to Gazzaniga (1989, 951), persons with a split brain "enjoy what appears to be a unified and unitary experience of conscious awareness". Similarly, Pinto et al. (2017a) state that while hemispherectomy patients are unable to integrate some information from both visual fields, and, in this sense, we can say that a split brain "divides" visual perception, this does not create two independent entities or two conscious observers.[39] Thus, when Parfit writes that "[t]he answer cannot be that these experiences are being had by the same person" (2006, 97), one possibility is that it actually can since a single conscious subject can experience two parallel and non-integrated streams of information (Pinto et al. 2017b).[40] Certainly, for a reductionist philosopher, this argument may not be a reason to correct the original hypothesis. On the contrary, it may be a reason to question the first-person account, because if the "split nature of the self", or the existence of two observers, is not subjectively felt and does not manifest itself in a person's behaviour, then the first-person perspective cannot be accepted as reliable. Of course, this might not yet be a problem, since there is a fairly long tradition in the history of philosophy of questioning intuitions and common-sense truths. Rather, a problem arises when one relies on notions and intuitions that are typical for the first-person perspective but when one's ontological and methodological approach is criterial (based on a third-person view, as Maria Schechtman would say).[41] Actually, there is no choice in the matter if one wants to remain intelligible, given that a third-person account that wanted to avoid the circularity objection would sound very strange indeed. From a first-person perspective, I can say "I'm getting married tomorrow." However, if I instead describe the situation as "A person in the future who is psychologically continuous with me is getting married," this is not a third-person statement but rather a confusion of the two perspectives, just as it is with the sentences "My replica thinks that he is me" and "He seems to remember my

[39] According to Pinto et al. (2017a), patients without a corpus callosum were able to "respond accurately to stimuli appearing anywhere in the visual field, regardless of whether they responded verbally, with the left or the right hand –despite not being able to compare stimuli between visual half-fields, and despite finding separate levels of performance in each visual half-field for labelling or matching stimuli". Moreover, and this seems to be the essential point in the whole matter, at the moment when the communication of information to the outside world was to occur, "the outcomes of perceptual processes are unified in consciousness, verbalization, and control of the body."

[40] For arguments in favour of the phenomenal unity thesis, also see Bayne and Chalmers (2003). Roland Puccetti (1981) argues that split-brain patients can be selves, or persons with two minds, where one can be unaware of what is happening in the other one.

[41] For a more detailed view, see Schechtman 1990.

life" (Parfit 1984, 219). This could appear to be a purely linguistic problem. I suspect, however, that it is not, since the third-person description here is pretending to be something that it does not possess and that it cannot provide from its level of argumentation.

In the next section, I will go back in time as I attempt to show how the beginnings of the reductionist approach to PI appear in Locke's *Essay*, which is considered to be the *locus classicus* of psychological theories of PI. In particular, I will note those places where Locke attempts to deny the relevance of the substantive self for the notion of identity and where he conflates different levels of argumentation for the purposes of the theory. I will also try to show why Locke's theory is a narrowing or a distortion of the relation between identity and memory and why circularity is not necessarily a defect but rather a sign of this distortion.

Persons with no past

Locke's theory of the person, despite easily identifiable points tempting one to confuse identity with continuity, differs from the neo-Lockean theories in one respect. Locke would probably disagree with the interpretation that persons responsible for past acts are "later parts of causal chains" or that they can survive if they retain at least half of their psychological connections to their yesterday's selves. In Chapter 27 of *An Essay Concerning Human Understanding*, Locke writes, "For as far as any intelligent being can repeat the idea of any past action with the same consciousness it had of it at first, and with the same consciousness it has of any present action; so far it is the same personal self."[42] Here Locke explicitly refers to "the same consciousness". But despite these seemingly clear references to "the same consciousness" – a consciousness more reminiscent of the substantive self of Joseph Butler and Thomas Reid, albeit presenting itself as something *distinct* – Locke's emphasis on identity cannot be taken very seriously.[43] What is this distinction? And why is it important that it can give the appearance of being an ontological distinction?

Let us begin with what has always been most provocative about Locke's theory, namely the insistence that it makes sense to speak of the same person across time only so far as his consciousness, namely the consciousness of past actions, extends (E II, xxviii, 10). According to the traditional interpretation of Locke's memory criterion, person *X* at time *t* and person *Y* at time *t1* are the same person if and only if *Y* at *t1* remembers from a first-person perspective what *X* was

[42] Also see E II, xxvii, 23: "So that self is not determined by identity or diversity of substance, which it cannot be sure of, but only by identity of consciousness."

[43] At least due to the fact that, unlike identity, continuity is not a transitive relation.

doing or experiencing at time *t*. To get a better idea of what "first-personally" means here, let us start with a little-known episode in Locke's life mentioned by Peter King.

On 15 January 1676, Locke noted in his travel diary that during a visit to Montpellier, he had bought twelve orange and lemon trees from a Genoese man at one livre a piece (King 1829, 55). If we were to ask Locke what it is that makes the person in the Montpellier market identical with the person who makes an entry in his diary on 15 January, he would probably answer that it is the awareness of past actions. If Locke remembers buying orange trees at Montpellier, or even sailing through the waters of the Flood on Noah's Ark (E II, xxvii, 16), then he is the same person as the person from Montpellier or from Noah's Ark.

Let us now look at a slightly different case. Let us suppose, however unlikely it may be, that Locke had not bought the goods in Montpellier but had actually stolen them and had shortly afterwards sustained an accident which caused him to permanently lose his memory of all previous events. According to the memory criterion, Locke before and after the accident would have been two persons rather than one. It might therefore be more correct not to refer to the later person as John Locke, despite a diary full of autobiographical notes written in his own handwriting or the personal correspondence addressed to Locke on his desk. This means, among other things, that the person after the incident (let us call him "John Locke") is no more responsible for Locke's actions than anyone else and that to punish him would be a wrong act. (I deliberately used the example of permanent retrograde amnesia, because common gaps in memory continuity, just like in Thomas Reid's brave officer paradox, could be solved in the ways suggested by Derek Parfit, Don Garrett, and John Perry). Naturally, Locke is aware that we would normally struggle to find something that would argue in favour of the identity of both persons, such as identical character traits, beliefs, desires, etc. However, the problem is that merely the knowledge that both persons have identical characteristics and patterns of behaviour is insufficient here, just as is a third-person knowledge of their past. Even if it might be interesting to learn something about "one's own past" from the telling of others, personal correspondence, and medical records, from a first-person perspective such a thing would be no different from listening to other people's biographies. The assumption that "John Locke" would be able to remember under normal circumstances also seems unhelpful, since it is ethically irrelevant if my memories are stored in intact inaccessible engrams or in God's mind, or whether I suffer from an extinction of memories (storage deficit) or the inability to remember them (retrieval deficit). If Locke were alive today, he would justify it somewhat like this: suppose person *S* commits a crime, say, robbing a bank, and then undergoes a medical procedure similar to the one described by Bernard Williams in *The Self and The Future* – he exchanges his body with that of an unknown person. The face of that unknown man would be on all the CCTV footage, so the person being punished for the robbery would be someone who has no memory of ever committing this act. If we agree that this is a case of injustice, we should interpret the theft of the orange trees in Montpellier in the same way or, as Locke writes (E II, xxvii, 26), any punishment for an act "done in another life, whereof he could be made to have no consciousness at all". Naturally, Locke knows that his approach might strike some readers as strange, and that this strangeness is due to the fact that in matters of moral or legal responsibility we do not rely exclusively on the first-person viewpoint. Rather, we gather evidence, take fingerprints, compare DNA, and rely on other people's testimony. If Locke were right, we could never say with certainty from a third-person perspective that we had the right person. However, the reason why Locke's account of the situation strikes us as strange lies in something else, namely in that it allows for the existence of very strange beings. Because if memory is what constitutes a person, then there are necessarily persons without a past – persons who have suddenly appeared (not knowing from where) in bodies that a short while ago had belonged to other persons. And about *their* future fate we know just as little.

Let us begin by stating that "John Locke" is not a person like others; after all, persons do have a past and autobiographical and semantic memories attached to it. Nevertheless, he is most certainly a person, or, as Locke himself writes, "a thinking intelligent being, that has reason and reflection" (E II, xxvvii, 9), and, moreover, a person who can perceive his situation as a loss. "John Locke" knows that he has lost access to a past that would help him clarify his current situation. The author of the *Essay*, however, sacrifices this past for the purposes of the theory, arguing that the idea of the pre-existence of amnesiac persons does not really refer to persons but to bodies with which these persons are mistakenly identified: "John Locke" does not have a past; it is only a body (a human) in which he is currently situated and which once belonged to another person, who has a past (E II, xxvii, 20). However, Locke does not seem to be very convincing here, since "John Locke" is not, after all, "a spirit wholly stripped of all its memory or consciousness of past actions" (E II, xxvii, 25). He may have, as is often the case in such situations, non-autobiographical memories of past events. However, even if he possessed no memories at all, which is an extremely rare occurrence, he would still possess other characteristics that are relevant to Locke's definition of a person. Thirdly, "John Locke" inherited his characteristics, beliefs, physical body, and social status from John Locke himself. Of course, persons are not bodies; nonetheless, they fulfil their intentions and desires through them. In this sense, bodies and physical consequences are the traces of persons' past lives. Suppose an action of mine causes a series of events but that the result of my action (e.g. an addiction, accident, illness, loss of property, or injury) will not be perceivable to me as a consequence of this action. The author of the *Viṣṇu Dharmasūtra* puts it a bit more poetically when he writes in VDS 20.47 that "just as a calf finds its own mother among thousands of cows, so actions done in a former life unerringly reach the perpetrator thereof" (Kane 1953, 40). If Locke wished to call such a case an injustice, then he would actually be conceding that causal laws – and, if there is a moral universe, also the laws of the moral universe – cannot operate beyond the ordinary concept of justice. However, these are already amplifications reaching far beyond the basic framework of Locke's theory. Moreover, they are not necessary when calling into question the metaphysical ambitions of Locke's concept of the person, because these ambitions encounter much more obvious obstacles. For example, there is the fact that persons, as Locke conceives them, become (or cease to be) persons in a gradual manner. This gradual aspect is intended to point to the blurring or problematic nature of the boundaries separating numerically distinct persons or persons and substances. The problem of boundaries presents itself most clearly in situations or cases that are usually described as "marginal", such as in young children, senile people, those suffering mental disorders, the insane, drunkards, and especially in long-term cases of amnesia with a happy ending. I will focus on this last case.

Imagine a situation – albeit a very unlikely one in everyday life – where a patient with long-term memory loss suddenly remembers and his mind fills with long-forgotten images. Squire et al. (1981) reported on the case of patients who, months after undergoing ECT, were able to recall their forgotten autobiographical past. How would such a case be interpreted in the context of Locke's theory? According to one definition (E II, xvii, 20), the moment of memory loss would separate two numerically distinct persons (*X* and *Y*), whereas the moment of recollection would mark the return of person *X*. In *Person and Object*, Roderick Chisholm expresses this using the following example: suppose I have to undergo a difficult and painful operation, but beforehand I have the opportunity to take a medicine which will make me forget my whole life thus far – who I am, how I got here, and so on. It is essential that I do not feel any pain, and since the self is constituted by consciousness and memories, the pain will be felt by someone else. After the operation I will take the pill again, after which time I will remember and then forget the painful operation. Chisholm asks: Where was I the whole time? Did I exist? How is it possible that I am back? Who was that other person? And where is he now? (1976, 110–111). For Chisholm, what Locke's theory cannot explain is long-term amnesia with a happy ending. However, there is another thing that is important to mention. Let us assume that the existence of temporary amnesia is an empirical fact documented by numerous neuropsychological studies. According to Locke's theory, we could express a case of amnesia as *J. L. t1* / "*J. L.*" *t2*. After recollection, more precisely the return of *J. L.* at time *t3*, we could interpret the situation as *J. L. t1* / "*J. L.*" *t2* / *J. L. t3*. Chisholm's question was: Who was that person "John Locke" at time *t2* and where is he now? The funny thing, however, is that *J. L. t3* autobiographically remembers "*J. L.*" *t2* and therefore must be the same person as "*J. L.*" *t2*, who failed to remember *J. L. t1*. The most plausible explanation for the whole story would be to concede that there never was any "John Locke", this strange person with no past, but only a John Locke before and after the accident. But if this is the case, then a person's persistence does not lie solely or necessarily in the current ability to remember, and the absence of memories is not a necessary criterion of PI. For this reason, neo-Lockean philosophers do not limit the psychological criterion exclusively to M-continuity; they replace it with a causal R-relation involving things like beliefs, intentions, and preferences, even though in this case the attempt to reduce identity to continuity may be problematic. Suppose a worst-case scenario were to occur and John Locke were to suffer an accident in which he lost all autobiographical memories and had his character, interests, beliefs, and the like significantly altered, as in the curious case of Phineas Gage as reported by John Harlow. Shoemaker and Parfit would probably warn us that to speak of "the same person" in this case is an offence against the notion of identity. Upon what basis could anyone claim that this is still the same person? In the following section, I will try to present two possible answers to this question and consider how plausible it would be to admit moral responsibility and practical concerns even in the absence of M-continuity.

Subjects and experiences

It may seem that if we take away the episodic memories of persons and much of what distinguishes them from others and by which they are identified, we will be left with only two options: (i) substance approaches or (ii) non-identity. However, in *The Phenomenal Self* (2008), Barry Dainton, a philosopher subscribing to the

Lockean tradition, argues that there is a third possibility. When it comes to the question of the persistence of the person, he argues that instead of the usual psychological criteria, a better strategy is to rely on something that is much more intimately and almost constantly connected to the person, and that is consciousness itself, or, more precisely, the continuity of consciousness. But what does the continuity of consciousness mean here? We can recall Williams' famous story of the mad surgeon and his involuntary patients, who feared a painful experiment despite assurances that they themselves would not experience any pain since their psychological profiles would be exchanged for those of other persons. Dainton asserts that if the surgeon's reassurances failed to allay their fears, it was not because they identified with the bodies but was rather because the continuity of consciousness can exist even with radical changes in the psychological spectrum that would be fatal according to P-theory. Parfit (1984) illustrates the impossibility of survival in the absence of corporeal and psychological continuity with a combined-spectrum thought experiment, which he uses to support the thesis that a person is not something simple in the sense of being "all or nothing" since there are possible cases where there is no true answer to the question of whether or not they are the same person. Since Parfit intended to use this example as an argument against the simple view, which I defend in this text, I will devote a few lines to it.

Let us imagine that I am kidnapped on the way home from a conference by some mad surgeon and that he performs a series of experiments on me. Firstly, he removes a small part of my memories and character traits and replaces them with Greta Garbo's memories and traits (Parfit's example). Then he replaces half of my memories with them, and ultimately then takes away all of my original memories and traits and replaces them with the memories and psychological profile of Garbo. According to Parfit, an adherent of the bodily theory of PI might argue that I had survived complete mind replacement since what matters is bodily continuity. And so, in the next version of the story (the physical spectrum), the surgeon gradually replaces all of my cells with those of Garbo. The problem is that this time my survival could be defended by a proponent of the psychological theory: I survived despite the fact that I have a very different body. The mad surgeon thus takes the final step: a replacement within both spectra with the result that the person at the end of the experiment will be physically and psychologically indistinguishable from Garbo herself. Parfit thinks that since there is neither psychological nor bodily continuity between me and the resulting person, no one would seriously claim that the person at the beginning and end of the spectrum is numerically the same person.[44]

If we note the particular direction of Parfit's argument, we find that both spectra and perspectives are combined in his example. First and foremost, it is not clear why, in the case of the combined spectrum, he does not argue equally as in the other two, say, by objecting that small changes do not matter and that at any moment I am the one who is experiencing the pain. Parfit is likely not wrong when he claims that it is impossible to imagine that the one who will experience pain in the middle of the experiment will only partly be me (1984, 233). Actually, for the same reason, I cannot answer the question "Is it still me?" other than in the affirmative. But if Parfit admits this monadic character of the self, at least for the purposes of the experiment, then why does he not argue in the same manner at the end of the experiment? Why does he not concede that if I can survive a change in one spectrum or the other, I can think of a continuous self even when both spectra change? One possible answer is that Parfit assumes that PI cannot consist in anything other than bodily or psychological continuity. Although in the first two versions he relies on first-person intuitions, that is, on a self that I cannot be only partially, in the third and final stage he lets this self disappears without a trace in the third-person perspective of "another person".

Let us now return to Dainton, who would do exactly what proponents of the "simple view" would do; he concedes that a person can survive radical changes in both spectra, but instead of "the same self" he will argue for the continuity of consciousness. John Locke could survive the loss of his memories, because he is phenomenally continuous with the John Locke that existed before the accident. This is not so simple, however, since without assuming the same subject of experience or the same stream of consciousness, we cannot reliably assess whether a person's mental states before and after the accident are linked with an uninterrupted stream of consciousness. How am I supposed to know that mental states separated in time are "co-streamal" and thus "consubjective" (2008, 379), being therefore part of a single stream of consciousness?[45] The problem becomes even more pressing when we try to defend the consubjectivness of experiences in the face of significant gaps in consciousness caused by dreamless sleep, seizures,

---

[44] Incidentally, this is precisely what Richard Swinburne (2013, 163) argues when he writes that it is metaphysically possible "that that substance acquires a totally new body, totally new apparent memories and character."

[45] Dan Zahavi (2011, 327) argues for a different approach when he writes that "the identity of the self is defined in terms of givenness rather than in terms of temporal continuity. Whether two temporally distinct experiences are mine or not depends on whether they are characterized by the same first-personal self-givenness; it is not a question of whether they are part of an uninterrupted stream of consciousness."

short-term losses of consciousness, amnesia, and so on. Consistent with the "default view", let us assume that in these situations it makes little sense to speak of a continuity of consciousness (for arguments in favour of the opposite view, see Thompson 2015). Dainton is well aware of these difficulties, which is why he ends up using the notion of "experiential powers" or the "capacity to have conscious states" to explain how the self can persist even in non-experiential phases. In other words, *x* and *y* would be phenomenally connected if certain experiences or mental states existing in the experiential powers were active; however, given the present state of knowledge, we cannot say exactly how these "powers" work or how they integrate, activate, form, or help to form the diachronic unity of consciousness. Nonetheless, it is precisely with this step – by replacing the criteria of actual continuity with the criterion of dispositional continuity – that Dainton moves away from an "phenomenalistic" perspective of the first person, which is a step that Locke could not afford to take. What brings him closer to Locke, however, is his uncompromising effort to establish the notion of persistence within consciousness, since consciousness has an epistemological primacy over the substantive self which we know little about. We do know, however, that it exists, since the concept of the non-self is phenomenologically extremely implausible: "If we are conscious, we can be certain that we exist, as subjects. What we cannot be certain about is what kind of subject we are" (2008, 254). For the purposes of this text, it is far more interesting to see whether or not this epistemological primacy of consciousness will – as in the case of Locke's memory – be elevated to an ontological criterion and whether Dainton would be tempted to speak of, say, a self constituted by consciousness or a subject of experience ontologically dependent on experience. For if a phenomenal consciousness requires a bearer or a subject, which Dainton accepts, then phenomenal consciousness ontologically depends on the existence of the self. I do not think this would be the best way forward for three reasons. The first reason is the impossibility of resolving the problem of gaps in consciousness within the phenomenalistic position. The second reason is the unanswered question of why we assume – if we proceed solely on the criterion of phenomenal continuity – that there is (or should be) a continuation of this continuity, even after alleged interruptions of the stream of consciousness.[46] The third reason is the unclear relation between phenomenal continuity and the "unknown" self.

[46] As Katja Crone put it in Phenomenal Self-Identity Over Time: "From a theoretical standpoint, what is actually required for identifying the bridge problem as a problem? [...] How can it occur to somebody that there is something like the bridge problem at all if he hadn't already a particular notion of what constitutes personal persistence? To put it differently, saying that something is wrong with the

Each of the reasons mentioned is, I believe, a good one for preferring a non-reductionist approach. However, Locke chose precisely the opposite path, which ultimately led him to several theoretical difficulties. Why did Locke think that a person's identity should be constituted by consciousness alone, instead of, say, immaterial substance? One possible answer is that when we are talking about the self, what is most crucial appears to be that what we can say about that self on the basis of first-person experience – what we are aware of. We have no clear idea of immaterial substance or indeed of substance in general. If we use the term "substance", it is only in the sense of a substratum, a bearer of properties which we are convinced cannot subsist *sine re substante* (E II, xxiii, 2). Locke even states that if "the same immaterial substance" were what constituted PI, I could not be sure that I was the same person that I was yesterday, since there are no criteria by which I could identify the thinking substance as being the same at different times. On the other hand, however, we have a clear awareness of our own thoughts, wishes, feelings, and memories as things that constitute the content of our conscious experience. Locke means here that it makes little sense to think of the self as something that I cannot grasp, that is different from what it appears to be. If I were in fact something other than what I am to myself, I would not be me.

At first glance, this reason would seem quite sufficient to reject the relevance of substance for PI. However, this is not as simple as it seems. Suppose that our notion of *x* is very vague and that we only know that *x* is a substratum or bearer of properties. In this case, how can we know that the self of first-person experience, which Locke defines as a "thinking intelligent being, that has reason and reflection" (E II, xxvvii, 9), is not just this *x?* One possible answer is that, in terms of Locke's notion of the person, this is not at all relevant; it is not relevant what *x* is if it is not what I can be aware of, or "what I can recollect" (E II, xxvii, 24). Nonetheless, it appears that the ambitions of Locke's theory reach much further when he claims that *S* could not have done *y* if he was not aware of *y*. The problem is, however, that he can claim something like this if he knows what *S* is but not when something merely appears to be *S*. To put it another way, to say that *S* could not perform the action of *y* if he was not aware of *y* means elevating the first-person experiential approach (along with all the possible risks) to being an ontological criterion: anything that is not first-personally accessible to *S* is not ontologically relevant for *S*.

---

experientiality claim pure and simple is to say that persons normally do persist – even though their streams of consciousness suffer interruptions" (2012, 211).

Understandably, there is another reason why Locke's theory of PI can *prima facie* get along without the immaterial substance or substance altogether. If we note the particular way in which Locke refers to persons, we find that his persons or consciousnesses are substances rather than properties and modes, since otherwise they could hardly be thinking intelligent beings possessing properties and retaining an identity across time as something distinct from individual acts of consciousness (E II, xxvii, 13). They cannot be modes simply for the fact that modes as thoughts or actions do not think and act, even though Locke's confusion of the terms "self" and "consciousness" attempts to mask this distinction. "The same consciousness", which is one of the most puzzling things about Locke's theory, has thus factually assumed the place of substance. This subtle exchange, however, as Chisholm (1976, 108) notes, paradoxically results in two thinking things in the chair with it not being clear which of them is currently thinking. Chisholm asks: Is it me who is thinking, and not the thinking substance? Why is it then called a 'thinking substance'? Is the thinking substance thinking and not me? This is as absurd an assertion as stating that neither one of us is thinking. Are we both thinking? This is an unnecessary multiplication of thinkers.

Is there any reasonable explanation for all of these oddities in Locke's theory? In *Past Lives of John Locke* (2018, 464), I am inclined to accept as most plausible the answer that Locke's main intention was not to formulate a metaphysical theory but rather propose a concept of the person that would be compatible with some forensic, ethical, and theological ideas, including the idea of the Last Judgment. For this purpose, the psychological criterion seemed to him to be the most appropriate one. This is also why he did not consider the question of an immortal soul or thinking substance to be that important, since – and in this he was critical not only of René Descartes but also of Henry More and, in general, of all of Cambridge Platonism at the time – the idea of immortality without memories was, in his view, both ethically and practically useless. Since this is a subject that survives in neo-Lockean approaches to this day, admittedly with minor modifications, and which tends to be used as an argument against substance theories of the person, I will make some remarks on it in conclusion. My aim will be to show that M-continuity is not a necessary condition for ethical and practical concerns. Let us start from Leibniz's well-known example:

> Suppose that some individual could suddenly become King of China on condition, however, of forgetting what he had been, as though being born again, would it not amount to the same practically, or as far as the effects could be perceived, as if the individual were annihilated, and a king of China were the same instant created in his place? (Leibniz 1951, 340)

The essence of Leibniz's argument is the claim that person *S* at time *t* cannot be interested in person *N* at time *t+1* if he knows that there will be nothing at the time by which person *N* can relate subjectively, that is, in a first-person manner to *S*. In other words, the form of survival offered by Leibniz's example does not contain or guarantee that which matters in survival according to Locke and Leibniz. It is not the quality of person *N*'s life corresponding to the nature of his past actions, or even a life characterized by some semantic memories that might be of some value to *N*.

Let us assume for a moment that Leibniz was serious about his argument, and that he truly believed that his existence without the ability to remember a previous life autobiographically was comparable in virtually every respect to his total extinction. The model of multiple lives would only be of interest to Leibniz if the Chinese ruler, or, if you will, "Leibniz" in the body of the Chinese ruler, remembered Leibniz the philosopher and dozens of other reincarnations in a first-person manner. While I am not convinced that many in Leibniz's position would have stood for such a privilege, or that it would have been a model of practical and psychological beneficiality (on the contrary, many persons would probably have preferred merciful oblivion, although they do not have to be sceptical about long stories like Galen Strawson), I will nonetheless try to suggest some reasons in favour of the opposite stance. Consider, for example, the fact that many of our activities, projects, plans, and so on are long-term in nature. Many of them are undertaken because we assume that we will be the ones who will experience the results of our actions. (Let us suppose that the condition "I know it will be me" is a prerequisite of prudential concern.) Nevertheless, by our own efforts, we cannot ensure that when the fruits of our labours ripen, we will still be able to connect them to our own past. These may be ordinary situations from real life, but the loss of that connection can happen much earlier and quite unexpectedly. Brian Levine et al. (1998, 1955–1957) describe the case of a patient M. L. who suffered from retrograde amnesia following TBI. After reawakening, he did not recognize his own wife, children, or family relatives, and he could not recall any episode of his life. Suppose for a moment that M. L. had learnt before the unfortunate event that in exactly one month he would suffer a head injury that would cause him to lose his memories permanently and that M. L. owned a large amount of property and was simultaneously undergoing treatment for a serious illness.[47] Does M. L. have

[47] We can, of course, imagine a different version of the example: all of M. L.'s memories would be stored and later transplanted (if such a thing is even possible) into the brain of person *N*, who would be stripped of all original memories (the same could be applied in the case of a broader R-relation). However, only one person could survive. I think that M. L.'s decision would not be at all clear-cut, and if he

a rational reason to act by continuing treatment and not spending all his wealth and so on? Two reasons seem the most likely explanation for acting in such a case: (i) the fact that persons are not indifferent to whether they suffer or enjoy despite their ignorance of the past events that led to this (in this sense, the future self is in the same position, i.e. just as ignorant, as the present self) and (ii) the belief that these events can be causally related to their past actions. For sure, Locke might object to the misuse of the term "their" in view of the absence of the criterion of re-identification and the circular description of the situation. A correct description would be: "Someone will suffer. I know that it is in my power to prevent that suffering." But if Locke's interpretation is correct, it should make no difference to me in principle whether I place my future situation in the hands of fate or prefer the other option and find myself in the situation of that unknown person (e.g. the King of China). Likewise, I should not care in principle whether I permanently cease to exist or start a new episode with no memory of the previous one.[48] The latter case would, of course, require my active approach to it.

Conclusion

Nothing that has been said thus far is meant to undermine the importance of memory and the ability to anticipate future experiences for a coherent sense of the self. My intention was rather to show that a memory theory conditioning identity on the continuity of memories is a narrowing or distortion of the relationship between identity and memory. The main problem of memory theory is not its circularity (the latter is, I believe, inevitable) but rather the fact that it does not sufficiently reflect the phenomenological level of remembering and forgetting which would allow for a distinction between memories and the remembering subject. Using several examples of temporary amnesia, I have attempted to show why trying to circumvent this subject – the self that is persisting even in the absence of memories – is a risky or even impossible step. I have also tried to show why excluding substance (the substantive self) from the identity game is extremely problematic. For if it is true that (i) we have no clear idea of the "thinking substance",

opted for the second option (survival without memories), it would be because he is not convinced of the correctness of the memory or the broader psychological criterion.

48 The very fact that I can appreciate the second possibility calls into question Leibniz's (Locke's) thesis about the uselessness or impracticality of immortality without memories. As noted above, forgetting can appear to be a practical and useful condition in many cases, especially when a new episode is associated with new "identities", characters, roles, or interpersonal relations. It would be psychologically difficult, if not impossible, to continue a new chapter of life with old memories. On the other hand, the inability to subjectively connect the two episodes does not mean that there are not any objective connections between them, including the principle of merit that matters to Locke.

(ii) the thinking substance and the person share the property of thinking, and especially (iii) if in the whole of the *Essay* we do not find a satisfactory account of the relation between substance and person, then the assertion of the irrelevance of the immaterial substance in the question of PI must appear to be considerably implausible.

I have argued in the chapter for the relevance of first-person awareness of the self as a place where our sense of transtemporal identity originates. For reasons mentioned in the first part this self is interpreted as the ontological or substantive self. At the same time, I have tried to justify why this ontological self, irreducible to facts about our minds and bodies, is a plausible solution to the "bridge problem" and a suitable approach to the transtemporal identity of a person. However, I did not rely exclusively on the first-person perspective: firstly, by the very fact of crossing the experiential plane towards the metaphysical (the persisting substantive self) and secondly, by the fact that I do not think that the first-person account will suffice for a plausible concept of the person. Much of what is called the "first-person perspective" contains elements of the social world. After all, in ordinary life we are not just transcendental subjects but also parents, sons, lovers, creditors, debtors, and sometimes people losing their memories. It is, after all, these others, as Leibniz (1996, 236) writes, who are the bridges to our past and who can tell us about it and bear witness to it, even when we can no longer identify with them. In *Amnesia and the Self* (2012), the neuropsychologist Daniel Levitin described the case of a patient with retrograde amnesia caused by a brain tumour who had spent the last weeks of his life surrounded by loved ones, listening to stories from his own life. He knew that he was leaving, yet he tried to piece the stories together to see in them a glimmer of meaning and excitement as at the time he was a part of them.

Acknowledgement

This chapter was written at the Department of Philosophy and Applied Philosophy, Faculty of Arts, University of Ss. Cyril and Methodius in Trnava, as a part of VEGA project 1/0563/18.

Bibliography

Bayne, Tim, and David Chalmers. 2003. "What is the Unity of Consciousness?" In *The Unity of Consciousness: Binding, Integration, and Dissociation,* ed. by Axel Cleeremans, 23–58. Oxford: Oxford University Press.

Bodhi, Bhikkhu (trans.). 2020. *The Connected Discourses of the Buddha: A Translation of the Saṃyutta Nikāya.* Boston: Wisdom Publications.

Chakravarti, Arindam. 1992. "I Touch What I Saw." *Philosophy and Phenomenological Research* 52, 1: 103–116.

Chisholm, M. Roderick. 1976. *Person and Object: A Metaphysical Study*. London: George Allen and Unwin.

Collins, Steven. 1994. "What Are the Buddhists Doing When They Deny the Self?" In *Religion and Practical Reason*, ed. by Frank E. Reynolds and David Tracy, 59–85. New York: State University of New York Press.

Crone, Katja. 2021. "Phenomenal Self–Identity Over Time." *Grazer Philosophische Studien* 84, 1: 201–216. DOI: 10.1163/9789401207904_010.

Dainton, Barry. 2008. *The Phenomenal Self.* Oxford: Oxford University Press.

Dainton, Barry, and Tim Bayne. 2005. "Consciousness as a Guide to Personal Persistence." *Australasian Journal of Philosophy 83,* 4: 549–571. DOI: https://doi.org/10.1080/00048400500338856.

Dennett, Daniel C. 2007. "Heterophenomenology Reconsidered." *Phenomenology and the Cognitive Sciences* 6, 1/2: 247–270. DOI: https://doi.org/10.1007/s11097-006-9044-9.

Dorahy, Martin J. et al. 2021. "The Sense of Self Over Time: Assessing Diachronicity in Dissociative Identity Disorder, Psychosis and Healthy Comparison Groups." *Frontiers in Psychology* 12: 620063. DOI: https://doi.org/10.3389/fpsyg.2021.620063.

Dravid, Narayan Shastri. 1995. *Ātmatattvaviveka of Udayanacarya with Translation, Explanation, and Analytical-Critical Survey*. Shimla: Indian Council of Philosophical Research.

Gazzaniga, Michael S. 1967. "The Split Brain in Man." *Scientific American* 217, 2: 24–29.

Gazzaniga, Michael S. 1989. "Organization of the Human Brain." *Science* 245, 4921: 947–952. DOI: http://doi.org/10.1126/science.2672334.

Grice, Paul. 1941. "Personal Identity." *Mind* 50, 200: 330–350.

Kane, Pandurang Vaman. 1953. *History of Dharmashastra*. Vol. IV. Poona: Bhandankar Oriental Research Institute.

Kant, Immanuel. [1781] 2007. *Critique of Pure Reason*. 2nd ed. Trans. by Norman K. Smith. London: Palgrave Macmillan.

King, Peter. 1829. *The Life of John Locke with Extracts of his Correspondence, Journals and Common–Place Books*. London: Henry Colburn.

Klein, Stanley B. 2012. "The Self and its Brain." *Social Cognition* 30, 4: 474–518. DOI: https://doi.org/10.1521/soco.2012.30.4.474.

Klein, Stanley B. 2014a. *The Two Selves.* Oxford: Oxford University Press.

Klein, Stanley B. 2014b. "Sameness and the Self: Philosophical and Psychological Considerations." *Frontiers in Psychology* 5, 29. https://doi.org/10.3389/fpsyg.2014.00029.

Klein, Stanley B. 2015. "The Feeling of Personal Ownership of One's Mental States: A Conceptual Argument and Empirical Evidence for an Essential, but Underappreciated, Mechanism of Mind." *Psychology of Consciousness: Theory, Research, and Practice* 2, 4: 355–376. DOI: https://doi.org/10.1037/cns0000052.

Leibniz, Gottfried W. 1951. *Selections*, ed. by Philip P. Wiener. New York: Charles Scribner's Sons.

Leibniz, Gottfried W. 1996. *New Essays on Human Understanding*, ed. by Peter Remnant and Jonathan Bennett. Cambridge: Cambridge University Press.

Levine, Brian et al. 1998. "Episodic Memory and the Self in a Case of Isolated Retrograde Amnesia." *Brain* 121, 10: 1951–1973. DOI: https://doi.org/10.1093/brain/121.10.1951.

Levitin, Daniel. 2012. "Amnesia and the Self that Remains when Memory is Lost." *The Atlantic*. December 31, 2012. Accessed July 4, 2021. https://www.theatlantic.com/health/archive/2012/12/amnesia-and-the-self-that-remains-when-memory-is-lost/266662/.

Locke, John. [1690] 1894. *An Essay Concerning Human Understanding*, ed. by Alexander C. Fraser. Oxford: Clarendon Press.

Madell, Geoffrey. 2015. *The Essence of the Self: In Defense of the Simple View of the Personal Identity*. New York, London: Routledge.

McDowell, John. 1997. "Reductionism and the First Person." In *Reading Parfit*, ed. by Jonathan Dancy, 230–250. Oxford: Blackwell Publishing.

McTaggart, John M. E. 1927. *The Nature of Existence*. Vol. II., ed. by Charlie D. Broad. Cambridge: Cambridge University Press.

Merricks, Trenton. 1998. "There Are No Criteria of Identity Over Time." *Noûs* 32, 1: 106–124. DOI: https://doi.org/10.1111/0029-4624.00091.

Minh, Châu. 1964. *Milindapañha and Nāgasenabhikshusūtra: A Comparative Study through Pāli and Chinese Sources*. Calcutta: Firma KL Mukhopadhyay.

Nagel, Thomas. 1971. "Brain Bisection and the Unity of Consciousness." *Synthese* 22: 396–413. DOI: https://doi.org/10.1007/BF00413435.

Ñāṇamoli, Bhikkhu (trans.). 2010. *The Path of Purification (Visuddhimagga) by Bhadantācarya Buddhaghosa*. Kandy: Buddhist Publication Society.

Ñāṇamoli, Bhikkhu, and Bhikkhu Bodhi (trans.). 2015. *The Middle Length Discourses of the Buddha: A Translation of the Majjhima Nikāya*. Boston: Wisdom Publications.

Parfit, Derek. 1984. *Reasons and Persons*. Oxford: Oxford University Press.

Parfit, Derek. [1987] 2016. "Divided Minds and the Nature of Persons." In *Science Fiction and Philosophy*, ed. by Susan Schneider, 91–98. Chichester: Wiley–Blackwell.

Pinto, Yair et al. 2017a. "Split Brain: Divided Perception but Undivided Consciousness." *Brain* 140, 5: 1231–1237. DOI: https://doi.org/10.1093/brain/aww358.

Pinto, Yair et al. 2017b. "The Split–Brain Phenomenon Revisited: A Single Conscious Agent with Split Perception." *Trends in Cognitive Sciences* 21, 11: 835–851. DOI: https://doi.org/10.1016/j.tics.2017.09.003.

Puccetti, Roland. 1981. "The Case for Mental Duality: Evidence from Split-Brain Data and Other Considerations." *Behavioral and Brain Sciences* 4, 1: 93–123. DOI: https://doi.org/10.1017/S0140525X00007755.

Rathbone, Clare J., Judi A. Ellis, Ian Baker, and Chris R. Butler. 2014. "Self, Memory, and Imagining the Future in a Case of Psychogenic Amnesia." *Neurocase* 21, 6: 727–737. DOI: https://doi.org/10.1080/13554794.2014.977923.

Rosenberg, Jay F. 1998. *Thinking Clearly about Death*. Indianapolis: Hackett Publishing.

Rozemberg, Andrej. 2018. "Past Lives of John Locke. Moral Responsibility and the Memory Criterion of Personal Identity." *Filozofia* 73, 6: 458–468.

Russell, Bertrand. [1913] 1992. *Theory of Knowledge,* ed. by Elizabeth R. Eames. London, New York: Routledge.

Schechtman, Maria. 1990. "Personhood and Personal Identity." *The Journal of Philosophy* 87, 2: 71–92. DOI: https://doi.org/10.2307/2026882.

Schwitzgebel, Eric. 2007. "The Unreliability of Naive Introspection." *The Philosophical Review* 117, 2: 245–273. DOI: https://doi.org/10.1215/00318108-2007-037.

Searle, John R. 2005. "The Self as a Problem in Philosophy and Neurobiology." In *The Lost Self: Pathologies of the Brain and Identity*, ed. by Todd E. Feinberg and Julian Paul Keenan, 7–20. Oxford: Oxford University Press.

Shoemaker, Sydney. 1994. "Self-Knowledge and Inner-Sense." *Philosophy and Phenomenological Research* 54, 2: 249–314. DOI: https://doi.org/10.2307/2108488.

Siderits, Mark, Evan Thompson, and Dan Zahavi (eds.). 2011. *Self, No Self? Perspectives from Analytical, Phenomenological, and Indian Traditions*. Oxford, New York: Oxford University Press.

Simeon, Daphne, and Jeffrey Abugel. 2006. *Feeling Unreal: Depersonalization Disorder and the Loss of the Self.* Oxford: Oxford University Press.

Sperry, Roger W. 1968. "Hemisphere Deconnection and Unity in Conscious Awareness." *American Psychologist* 23, 10: 723–733. DOI: https://doi.org/10.1037/h0026839.

Squire, Larry R., Pamela C. Slater, and Patricia L. Miller. 1981. "Retrograde Amnesia and Bilateral Electroconvulsive Therapy: Long-Term Follow-Up." *Archives of General Psychiatry* 38, 1: 89–95.

Strawson, Galen. 2017. *The Subject of Experience*. Oxford: Oxford University Press.

Swinburne, Richard. 2013. *Mind, Brain, and Free Will.* Oxford: Oxford University Press.

Taber, John A. 1990. "The Mīmāṃsā Theory of Self-Recognition." *Philosophy East and West* 40, 1: 36–57. DOI: https://doi.org/10.2307/1399548.

Thompson, Evan. 2015. *Waking, Dreaming, Being: Self and Consciousness in Neuroscience, Meditation, and Philosophy.* New York: Columbia University Press.

Tulving, Endel. 1993. "Self–Knowledge of an Amnesic Individual is Represented Abstractly." In *Advances in Social Cognition,* ed. by Thomas K. Srull and Robert S. Wyer, 147–156. Hillsdale, NJ: Lawrence Erlbaum Associates.

Williams, Bernard. 1970. "The Self and the Future." *The Philosophical Review* 79, 2: 161–180. DOI: https://doi.org/10.2307/2183946.

Zahavi, Dan. 2011. "Unity of Consciousness and the Problem of Self." In *The Oxford Handbook of the Self*, ed. by Shaun Gallagher, 316–338. Oxford: Oxford University Press.

Chapter 3
The Origin of Rules

Tomáš Čana

Abstract: The problem of the origin of the rules which we govern ourselves by was brought to the centre of attention by Ludwig Wittgenstein. In examining his legacy, four motives come into the spotlight. Firstly, man should be seen primarily as a being who acts and only secondarily as a being who is able to reflect upon his actions. Secondly, rule-following would be inconceivable without a certain uniformity in our inner experience. It would also be inconceivable without a uniformity in the external manifestation of our inner experience. Thirdly, rule-following would also be inconceivable if we had not completed training of sufficient duration in applying certain sounds, expressions, principles, metaphors, and suchlike. The fourth and final motive is based on the conviction that the previous points cannot constitute a basis for forming a theory. From understanding these points, we should be able to directly achieve an anti-theoretical position. While I do not have any issues with the first three points, I see multiple issues with the fourth one. I therefore agree with Wittgenstein's opinions on the genealogy of normativity, but I disagree with the conclusion he draws from them.
Keywords: Form of life, universal agreement among members of a species, normativity, training, upbringing, theory, acting, Wittgenstein.

The problem I want to address here is not an eternal philosophical question. (It is not older than Darwin's theory of the origin of species, without which it probably could not have been formulated.) The question about the origin of the rules, regulations, and commandments that affect us in everyday life was consciously posed for the first time by Friedrich Nietzsche in the nineteenth century. Nietzsche formulated this question almost exclusively in terms of the criteria according to which we make our decisions when making "ethical" and "aesthetic" judgments (Nietzsche 1997). In the twentieth century, the problem of the origin of functioning rules was brought back to the centre of attention by Ludwig Wittgenstein, who not only examined the origin of selected criteria – such as the criteria for good and bad – but also attempted to uncover the genealogy of functioning criteria in general. In other words, Wittgenstein was interested in the origin of concepts such as rule-following, errors, and rule-breaking. He ultimately asks: Where did normativity itself come from?

This chapter is based on the basic assumption that the abovementioned interest in the genealogy of normativity has brought many interesting and important results when shifted to the general level. At the same time, this is based on the assumption that these results have not yet been understood in a sufficiently clear manner – and not only in our cultural space. My aim is to firstly provide a clear reconstruction of the tangible results that Wittgenstein achieved with regard to this theoretical interest. My second aim is to take a critical stand towards them by distinguishing between those that are acceptable and those that are not.

1.

It should be noted at the outset that the question of the origin of functioning rules was a constant and lifelong source of interest for Wittgenstein. In his early, middle, and later creative periods, he remained interested in the roots of the rules, regulations, and commandments that really affect us. (Although, it is true that this interest manifested itself in a significantly different manner at different times in his life.)

It should also be noted that his posthumously published writings – in this case, I am referring especially to the later ones – enriched modern philosophical thought as such with the introduction of this problem. These writings became the subject of a particularly large number of critical discussions and fierce debates within both analytic and continental philosophy. They have also provided the basis for a significant number of mutually exclusive philosophical interpretations (e.g. Horwich 2012; Schneider 2014; Maddy 2014; Coliva 2015; Schönbaumsfeld 2016).

At first glance, Wittgenstein's later interest in the origin of normativity is characterized by a limitless variety of topics. In fact, he dedicated an equal portion of his attention to the criteria we use to navigate through the fields of psychology, mathematics, logic, arts, religion, and ethics. He analyses the normative factors in buying apples at a grocery store and in giving orders on a construction site. We could say that he tries to approach a given problem from as many points of view as possible. In the end, however, his goal remains unchanged. This is to answer the question: What are the rules that affect me in life based on? How is it possible that they are able to affect me? And, on the other hand, what allows me to ignore or refuse them under certain circumstances? In other words, what is the activity of governing ourselves by the rules that we observe around us (or rejecting them) based on? Can it be rationally understood and explained?

As I have already indicated, I will first try to reconstruct what our position is in connection to this problem according to the later Wittgenstein.

2.

When examining Wittgenstein's legacy in connection to the genealogy of normativity, four key points ultimately come to the fore:

1. the priority of acting over thinking
2. a universal agreement among members of a species as a first condition

3. the completion of a certain form of drill as the second condition
4. an anti-theoretical stance with respect to the "classical" theory of the origin of normativity.

I will further try to break these key points down into smaller units. However, it should already be mentioned that the first point represents a strong methodological recommendation for philosophers. According to Wittgenstein, without its acceptance we will never be able to achieve a correct view of the activity of governing ourselves by rules. The second and third points represent two attempts to provide concrete substantive answers to the main question of this investigation. These points should be understood as something complementary (like two episodes of a single story). Indeed, "What we are supplying are really remarks on the natural history of human beings; we are not contributing curiosities however, but observations which no one has doubted, but which have escaped remark only because they are always before our eyes" (Wittgenstein 1999, §415). The fourth point deals with the consequences that should result from the previous three points for something like a coherent theory of the origin of normativity.

3.

The first point represents a certain methodological recommendation. It should be understood as a reminder of a fact which we tend to forget about, especially when constructing thought experiments in philosophy.[49] The main idea is that if we want to avoid sceptical doubts and philosophical speculations that offer little hope of bringing something useful, we should view man primarily as a being that acts and only secondarily as a being that rationally reflects upon his actions. This means that Wittgenstein's writings portray the individual as someone who first acts in a specific – and not in any other – way and only later is capable of asking questions about this behaviour. First comes something like "this action", and only afterwards comes the possibility of asking questions about it and evaluating or doubting it. Just as Goethe said: "In the beginning was the deed" (Wittgenstein 1972, §402), and only on this basis was the Word born. Only subsequently – upon this background – does there then arise the possibility of constructing theoretical systems and explaining things rationally.

According to Wittgenstein, the Cartesian object that thinks is essentially made possible by the fact that something as a thing that acts in a specific (and not in any other) way has previously been established. The "pure reason" which deals

[49] This is a reference to hypotheses such as Descartes's evil deceiver (Descartes 2005, 3–62); Ayer's Robinson Crusoe, who tries to invent his own private language (Ayer 1966, 259–263); Kripke's metalinguistic sceptic who proposes bizarre – although not *a priori* excluded – interpretations of the sign "+" (Kripke 2002, 7–54); and Putnam's brain in a vat (Putnam 1981, 5–8).

with non-applied theoretical research exists only because we have previously encountered successful operations of practical (applied) reason time and time again. Abstract theoretical thinking is not a precondition for our everyday practice, as it might seem to be based on the reading of classical philosophical texts from Parmenides to Husserl, but rather its concomitant or consequence.

Wittgenstein, however, does not imply that thinking is not important, or that reason plays anything other than an exceptionally important role in our lives. He does not promote voluntarism, emotivism, or irrationalism, nor does he attempt to question our capability of planning rationally for the future. What he tries to say is that if we want to find a general perspective from which we would remain unmoved by the majority of traditional philosophical questions about rules, we should not forget that before we even start thinking about them, we are already acting on them. If we want to stand on solid ground, we should not forget that a specific (and no other) action always comes first, and that only afterwards do we think about this action (and anything else). According to Wittgenstein, in this regard, the correct order is crucial: "Doubting and non-doubting behaviour: there is the first only if there is the second" (1972, §354).

Wittgenstein argues that the fact that we are already acting in a particular way before we even start thinking about the rules implies another important fact: all around us, there is a certain practice at any given time. There are time-proven procedures and rituals, and with them come the criteria for applying terms like "right", "wrong", "true", "false", "normal", "abnormal", "moral", "immoral", and so on. Any reasoning and theorizing always develops within such a practice, and one cannot think without its contribution; or rather, one cannot think without paradoxes (Kripke 2002, 55–113). This means that the questions we ask and the theoretical hypotheses we place before others always result from a certain way of acting that we understand (Wright 2004a). Otherwise, they are impossible – whether we realize it or not.

4.

The second point represents more than just a methodological recommendation. We progress here from methodology towards problem-solving. Specifically, this point represents the first part of the answer to the main question of this study.

The activity of rule-following would be inconceivable without a certain uniformity in our inner experience, feeling, and sensory perception. This means that we resemble each other as members of a species in what we feel when experiencing joy, when we dislike something, when our knee hurts, and even when we perceive a specific tone or a characteristic smell. We are only speaking here of a uniformity in the "normal experience", feeling, and perception (although not in all cases). Wittgenstein reminds us that all people that are normal resemble each other in their inner processes. He adds to this postulate that if this were not the case – if we were all inimitable and unique – we could not encounter the activity of rule-following. There would be no way of – and no basis for – establishing this activity: Indeed, "In order to make a mistake, a man must already judge in conformity with mankind" (Wittgenstein 1972, §156).

This activity would be equally inconceivable without any uniformity in the external manifestation of our inner experience, feeling, and perception. This means that we also resemble each other as members of a species in the way of manifesting happiness externally, or in the impression we give when we dislike something, when our knee hurts, or when we perceive a certain tone or a characteristic smell. Again, we are only talking about uniformity in the "normal expression" of our experience, feeling, and perception (not in all cases). Wittgenstein reminds us that all people that are normal manifest their inner processes in a similar manner. He adds that if this were not the case – if there were no regularity in the physical manifestation of our inner processes – we could not encounter such a concept as normativity in the first place. There would be nothing objective upon the basis of which we could acquire it. Indeed, "What would it be like if human beings shewed no outward signs of pain (did not groan, grimace, etc.)? Then it would be impossible to teach a child the use of the word 'tooth-ache'" (Wittgenstein 1999, §257).

This leads one to the conclusion that the activity of rule-following (or rule-ignoring) that we observe in the world around us under normal circumstances assumes the existence of an agreement among members of a species. This agreement is observed on two levels:
(a) in the physiological reaction to what happens around us (and inside us) and
(b) in the characteristic physical manifestations of this physiological reaction.

Wittgenstein stresses that without accepting these two postulates, we will never be able to get to the bottom of this problem upon which everything is based. In fact, we have just reached the lowest level which the rules, regulations, and commandments that affect us in life grow from. No rule can affect anyone without some kind of consensus behind it. The terms "rule" and "consensus" are deeply interconnected and are like close relatives. In any case, the activity of rule-following cannot exist – nor can it be conceived – without there being the presumption of a consensus among people.

As an illustration, let us use greetings as an act of speech which every one of us has participated in. This act appears to be an elementary operation with symbols. According to Wittgenstein, however, the possibility of a greeting is based on the fulfilment of multiple conditions which we usually forget about. What conditions exactly? For example, conformity among speakers in the perception of a certain behaviour as being normal (and certain circumstances as common) and conformity in the perception of a certain demeanour as problematic (and certain circumstances as abnormal). This is based on the perception of numerous factors and aspects as being normal and others as being controversial or disturbing. After all, "If language is to be a means of communication there must be agreement not only in definitions but also (queer as this may sound) in judgments" (1999, §242). According to Wittgenstein, this is where the foundations of understanding between people lie. Without an elementary consensus on what is normal and what is disturbing – as some form of a step zero – a communication of a certain meaning, such as a greeting, would be impossible.

However, an important issue is that the conformity between people in their internal processes and the associated expressions must be universal. All people – and all cultures – should, under given circumstances, view this for what it is. In simple terms, all people that are normal are alike; or, all people that are normal are the same. Without accepting this postulate, says Wittgenstein, we cannot make any progress in the investigation of the origin of normativity.

What does this mean? Despite all the individual peculiarities we have encountered in the past and the differences that may come to mind in connection with other cultures, the things that we have in common as individuals and as cultures, and the things that we have which are similar, are far more numerous. Incomparably so. If this were not the case – if the things that are different and foreign prevailed over those that we have in common or that are similar to each other – it would not be possible for us to even greet each other. And this is not all.

It would also be impossible to transfer information from one culture to another; for instance, it would be impossible for different cultures to influence each other. (Although we know that successful information transfer between different cultures does occur.) From Wittgenstein's point of view, the concept of "information transfer" or "translation" necessarily presumes the notion of the common behaviour of mankind. Otherwise, these concepts cannot be understood or reconstructed. Indeed, "The common behaviour of mankind is the system of reference by means of which we interpret an unknown language" (1999, §206).

Another important notion in this framework is the fact that the idea of a universal agreement between members of a species does not refer to conventions we would all be able to agree upon. We are not speaking here of a social contract or a social institution that we have established in the past. According to Wittgenstein, we should not forget that the term "agreement" refers to what we have agreed upon as well as something categorically different. Its original meaning refers to something instinctive and rooted in ourselves: something that grows, figuratively speaking, from our physiology (Moyal-Sharrock 2017, 554–555). Indeed, this is something "[...] that lies beyond being justified or unjustified; as it were, as something animal" (Wittgenstein 1972, §359). So, what is it exactly? It is everything that we as members of a species have in common and that makes us the same as human beings.

Ultimately, this is nothing but a statement that we are such and such – and not any different. We react in one way and not in another. Some things we do perpetually and other things we do not even consider doing. We accept this while we are suspicious of that. Perhaps, as Wittgenstein sees it, this sounds trivial; however, when trying to understand the genealogy of criteria according to which we normally operate, we cannot ignore this postulate. More precisely, to ignore this reality most likely means to consequently lose our way (and end up on a Platonian or conventionalist wayward course). And why should we expect, while examining the genealogy of functioning criteria, anything other than triviality? Why should the fact that we are of a certain nature and not any different not lie there – in the deepest foundations of our procedures?

A question certainly comes to mind at this juncture: what are we actually like? We should not expect a definition as an answer (just as we are not able to define either "normality" or "abnormality"). Even if we cannot define our nature, we can at least make it clearer in a way. And how so? According to Wittgenstein, we can achieve this through an effective process of exemplification.

Our nature is such that if someone explains to us what colour "Parisian blue" is by pointing their finger, we will not even think of looking the other way. (We look in the direction of their finger.) When someone teaches us how to add numbers and happens to be using examples that do not exceed the number 1,000,000, we will not assume that anything would be different after passing that threshold. From the moment we have understood – on the basis of a specific explanation – what it means "to add", we add the numbers the same way regardless of a number's value. (It does not occur to us that the rules might change radically with higher numbers.) When somebody explains to us how the names for musical notes work on a certain day of the week (for example, Monday) we do not expect that that these names would function differently on another day (for example, Thursday). (Such things do not occur to us normally.) And so, it goes. This, however, does not mean that making such moves is irrational and stupid in itself. It means that under normal circumstances, these thinking patterns represent something that none of us would think of doing. (In this sense, we can rightfully define them as irrational and stupid.)

According to Wittgenstein, if there were a creative person who would think of these alternatives, it would firstly mean that they are completely different from us and secondly that it is unlikely that such a person would learn to use the names of musical notes or add numbers correctly. Why is this? Because such a person would not be able to find out where the demarcation line lies between what is correct and what is incorrect. Due to their unique creativity, they would not be able to learn this difference. This would be a problem. Their thinking patterns would consequently be perceived as disturbing (and we would most probably classify them as "irrational and stupid"). Wittgenstein also reminds us that if a situation should occur in which such ideas began to appear in the minds of a considerable number of the members of our species, it would mean that the practice of adding, pointing, or greeting could not be introduced into practice (Rheese 1966, 268–269). It could not be constituted, because it would be lacking a basis. Consequently, the existence of the objective thought content that we would be able to (and would want to) communicate to each other (either as individuals or in some other form) would also be impossible.

> It is only in normal cases that the use of a word is clearly prescribed; we know, are in no doubt, what to say in this or that case. The more abnormal the case, the more doubtful it becomes what we are to say. And if things were quite different from what they actually are – if there were for instance no characteristic expression of pain, of fear, of joy; if rule became exception and exception rule; or if both became phenomena of roughly equal frequency – this would make our normal language-games lose their point. (Wittgenstein 1999, §142)

For Wittgenstein, the core idea is that as people we are of a certain nature. Without a reflective, instinctive, and physiological conformity which connects us as members of a species, we would not be able to sit down at the table and agree on something (for example, a new convention). When we try to understand what is really going on when we govern ourselves by specific rules, concluded contracts and the resulting obligations are insufficient. We cannot operate with only what we have agreed upon. This is not enough. We need something categorically different to build upon: every aspect in which we are more similar and related to each other than in which we are different and foreign.

5.

The third point represents an attempt to propose the second part of the answer to the main question of this study. As has already been stated, given the content of the second point, the relationship between the second and third points is one of complementarity. Consequentially, it is not always obvious where the demarcation line between them lies. Nevertheless, being able to distinguish between them is important.

The activity of rule-following would be inconceivable without us all – at a specific stage in our lives – having completed training of an appropriate duration in applying certain sounds, gestures, postures, expressions, principles, metaphors, and so on. More precisely, this activity could not have taken place had we not been exposed repeatedly to external circumstances under which we were led to react to specific characteristic stimuli in a certain unique way. Just like that, without question, and without any further reasoning. If we had not been allowed to undergo such a process of conditioning – and also without our disposition of allowing ourselves to be "trained" in such a way – we would not have been able to work our way towards the notion of normativity (communication, values, upbringing, etc.). According to Wittgenstein, we would have never been able to understand what it means for something to have a meaning and what it means for something to symbolize something else. Also, we would not have been able to understand what it means to learn something new.

Paradoxical as it may sound, man's ability to acquire new information and skills is based on the primary development of his ability to carry out many operations mechanically. As if he were a machine. In addition to what we have in common as members of a species, our ability to learn is conditioned precisely by undergoing (at the right age) such a drill to the full extent. This means that on their own, the universal aspects in our physiological equipment are not enough for us to arrive at a functioning notion of "normativity". Even though, according to Wittgenstein, it is impossible to – metaphorically speaking – "push off from the bottom" without the animal aspect, something crucial (at a purely animal level) is nevertheless missing.

And what is missing here? What is missing is the obedience training we have all been through. Someone who would again and again repeat my name until I figure out how to use that sound. Someone who would endlessly repeat the question "What is this?" while they point their finger in front of me until I realize that everything I see around me is composed of units which I attach names to. What is missing is the praise that makes me happy and the reprimand that I want to avoid. What is therefore missing are other people; without their presence and interventions (such as praise and reprimands) the question of the origin of functioning rules would not stand on solid ground. From Wittgenstein's point of view, in such a case, this problem would be suspended in a vacuum. In order to avoid this traditional fate of philosophical analyses, our model needs to include other human beings who we observe and imitate (or even reject), and thanks to whom the possibility of one's own formation as a human being is created.

Why does this matter? When we take a look at ourselves when inferring or calculating, we cannot, according to Wittgenstein, overlook the fact that aside from the components that are worthy of retrospective reconstruction as the activities of interpretation of sense; we also encounter categorically different components that lack any such association. This means that what we see here are actions that we perform correctly without anything like a conscious interpretation or an explanation hiding behind them. It is not true that everything we refer to as "the activity of rule-following" must contain the activity of the interpretation of sense as a specific connective component. Even though Wittgenstein himself stood behind this theory in his early writings (1971), this is not the case. In fact, there is nothing behind many of our "rational actions" other than a successful process of conditioning.

When we focus on ourselves while proceeding in accordance with a basic law of logic, such as the law of separating a conjunction (according to which p and q being true implies that p is true), we realize that we perform this action similarly to the way that a lightbulb is turned on by a switch. We turn the switch on, and the lightbulb lights up. We turn it off, and the light goes out. According to Wittgenstein, this is not an activity of interpreting a meaning – and definitely not an activity of reception – as there is no interpretation to be found. It is a mechanical step we take because we are trained to do so, because we have developed a specific mental mechanism. Indeed, "The drill of teaching could in this case be said to have built up a psychical mechanism" (Wittgenstein 1958, 12). If someone were to ask for the reasoning behind this action, we would probably not perceive them as an objective critic but rather as someone who undermines our elementary competence (Wright 2004b, 163–164).

According to Wittgenstein, multiple components in the framework of operations that we perceive as rational are nothing other than the act of actual rule-following. On the contrary, multiple components that we perceive as irrational and stupid are nothing more than an actual disregard of the rules. There is nothing else behind it. Indeed, "What this shews is that there is a way of grasping a rule which is not an interpretation, but which is exhibited in what we call 'obeying the rule' and 'going against it' in actual cases" (Wittgenstein 1999, §201). What does this mean? It is simply what we refer to as calculating or counting and what we refer to as valid inference or deduction. There is no deeper reasoning, nor can one be discovered.

This leads us to the conclusion that simply understanding the constitutive components of our procedures in which we participate together with other normal people is not enough. I need something more than just someone explaining to me in detail how the rule of separating or introducing a conjunction works. That is not enough. My personal history must contain a reasonably long process of conditioning, thanks to which I see no other option than to process the information from other people (and then give it back) in accordance with these norms and without any form of interpretation. In fact, according to Wittgenstein, nobody is even required to explicitly state these rules. I do not need to know about their existence (just as the majority of speakers have no idea about the existence of something like a basic logical rule). However, I need to be correctly trained in terms of how they are commonly used, and this training must take place at the right age. Thanks to the training I have undergone at a certain age, the only possibility for me is to react in this way: in accordance with the rule of separation of a conjunction and without thinking. It is done unconsciously (in the sense of consciously meaning being too slow).

From stating that simply understanding the constitutive components of our procedures is not enough, an important general conclusion can be reached. According to Wittgenstein, it can be said of every human being with whom we have ever successfully communicated – for example, when greeting each other – that at a specific stage of their life they were willing to receive a considerable number of instructions without questioning them. They were willing to accept many statements from other people without questioning them, doing so as if it was automatic. Every being that we can correctly characterize as "communicating" has inevitably, at a certain stage of their life, undergone some form of obedience training. There is no other option for consideration.

The truth is that this does not sound very flattering. Nevertheless, this statement is crucial with regard to the research of the origin of normativity. At the same time, this results in a variety of important individual consequences. According to Wittgenstein, we should, for example, not forget that without the willingness to accept certain instructions without any doubts (just like in obedience training) there is no way for us to acquire our mother tongue. In such situations – without the "primary language" – it is impossible to make any progress with regard to the research of the origin of normativity (as well as any other research). Without any drill-like element contained in the personal history of each of us, there can be literally "no speaking of" rule-following or of rejecting rules. To the contrary, such terms would become meaningless and would lose all objective meaning.

Finally, it is necessary to add that in connection to the content of both the second and third points as a whole, Wittgenstein sometimes speaks simplistically of the form of life. He speaks of something concerning the animal (instinctive) experiencing upon which we agree, but only under the presumption that we have undergone training of a sufficient duration in the use of certain sounds, gestures, poses, expressions, principles, metaphors, and so on.

6.

The fourth point concerns Wittgenstein's attempt at a critical reflection on everything that has been said in connection with the main research question. This critical reflection leads to a strict anti-theoretical stance towards the presented theoretical problem.

The fourth point presents the conviction that on the basis of the previous points – (1) the priority of practice over reflection, (2) uniformity in our normal experience and expression, and (3) obedience training of a sufficient duration – it is impossible to construct anything like a "classical" theory. On the contrary, the rational reconstruction of the genealogy of functioning rules should appear to be both pragmatically and logically impossible after a more careful observation. Wittgenstein therefore reaches the conclusion that, generally speaking, this inseparable part of our usual functioning cannot be understood. Nor can it be rationally explained. Why is this? There are at least two reasons.

The previous three points – should someone decide to give them a systemic form – primarily represent something too inconsistent, heterogenous, and indeterminate. From Wittgenstein's point of view, the effort to connect them as components of a single model does not make any sense. How would we even want to connect such different things? In this regard, he reminds us that the contents of the second and third points necessarily include facts which are difficult to explain in principle. For instance, they include that which is not talked about under normal circumstances but that which is shown: these are the aspects which we agree on the acceptance of without the need of "knowing" about them explicitly (or having their definition).

Wittgenstein asks: how would we want to incorporate, for example, an ostensive definition of meaning, which must include our correct view, into a theory? How would we want to make a correct way of pointing at something the subject of interpretation? And how would we want to define concrete circumstances which we are exposed to at a specific stage of our lives? How far can we get in terms of such an effort? Trying to achieve a rational explanation in this regard is the manifestation of our pseudo-rationalistic tendencies. This effort is, in fact, the manifestation of a deficiency in understanding. And we should not follow through with it. An explanation in this case is something that we – if we see things correctly – do not need. Indeed, "What is 'learning a rule'? – *This*. What is 'making a mistake in applying it'? – *This*. And what is pointed to here is something indeterminate" (Wittgenstein 1972, §28).

In this context, however, we encounter the obstacle of circularity. This is the second reason Wittgenstein refuses the possibility of a theory of the origin of normativity. From his perspective, from the moment we start reflecting on the activity of rule-following – and later, when we try to talk about it – we are already in a circle. Why is this? Because in terms of this effort, we are unable to move without the rules by which we (once again) govern ourselves. In connection with the activity of rule-following, there is therefore no opportunity for us to discover anything that would resemble a neutral point of view. On the contrary, and in retrospect in this investigation, we are akin to the animal that bit itself in Nietzsche's well-known parable (Nietzsche 1968). In any case, we should be aware of the fact that from the beginning – from the first point to the third – we have been in a circle. More precisely, we have based our clarification of the previous points on precisely the thing which we have also been trying to understand the roots of.

Is it possible to overcome this difficulty? According to Wittgenstein, no. Firstly, there is no logically possible world in which we could speak and yet take a neutral standpoint with regard to the activity of rule-following. In other words, there is no possibility to make a universally valid statement about rule-following while simultaneously ignoring specific established rules. No such thing can be done without falling into circularity. Secondly, even if in some extraordinary way we discovered such a neutral standpoint, we would have to remain silent. Why is this? Because at the same time, we would lose all the obvious criteria we normally use to navigate in the world and in ourselves (Davidson 2006). We would, in relation to our notions and categories, lose the solid ground from under our feet.

These are the two arguments for the fourth point. They are also the two reasons why, according to Wittgenstein, we should not aspire to construct a coherent theory of the origin of rules.

7.

Over the following part of this study, we will move on from reconstructing the results of Wittgenstein's interest towards criticizing them. In relation to that, I state that while I have no problem with the first three points, I see multiple difficulties arising from the fourth one. In essence, I agree with Wittgenstein's opinions on the genealogy of normativity, but I disagree with what he believes they should imply.

I agree that if we want to avoid purposeless discussions about rules, we should see man primarily as a being that acts and only secondarily as a being that can rationally reflect upon this action. I also tend to agree with the statement that the activity of rule-following that we see around us is conditioned by a universal agreement among the members of a species (that is, by the similarities in our normal experiencing and expressing) as well as by a "drill" that all of us (who communicate successfully) have undergone at a specific stage of our lives. This means that I accept Wittgenstein's methodological advice as well as both of the attempts he makes at a substantive response to the main question of this investigation. (At the same time, I am convinced that thanks to these exact results, we will be able to acquire a much more advantageous position during a future investigation of this problem compared to what we had in the past.)

On the other hand, I do have a problem with the persistent refusal of the possibility of constructing a coherent interpretation of these results that is also present in Wittgenstein's work. According to Wittgenstein, we should be able to move from the first three points directly to the anti-theoretical position (quietism). From his perspective, there is no other possibility. Personally, however, I do not consider the inclination to anti-theorism or quietism to be essential. I understand the reasons that lead Wittgenstein to the refusal of a rational explanation of the genealogy of normativity, but I do not agree with them. On the contrary, I consider them to be rushed and overly pessimistic.

In connection to this, it should most probably be noted that it is precisely the fourth point that turned out to be the particularly controversial one in philosophical circles. For a number of thinkers, Wittgenstein's conclusion clearly represents something they cannot live with (Boghossian 2000, 234). The reasons supposedly leading to it are therefore repeatedly subjected to criticism. Here I present two representative illustrations upon which I will build my own critical evaluation of the fourth point:

In "The Justification of Deduction", Michael Dummett draws attention to the fact that circularity should not be the cause for any such uproar (Dummett 1978). We should not forget that there are contexts in which the "movement in circles" does not pose any difficulties. But what contexts are these? For example, this can be when, with the help of a deductively valid argument, we attempt to propose an explanation of a conviction that we all agree upon. It can be, for example, the explanation of the conviction that "ice floats on water" in such a way that we create supplementary premises. If something has a specific weight that is less than that of water, it will float; and ice has a specific weight that is less than that of water.

Dummett admits that an argument which is constructed in such a way is circular. This means that the conclusion in its entirety is already included in its premises. At the same time, he is convinced that this fact does not pose (under these circumstances) any problem. On the contrary, circularity is completely uninteresting in this case. Why is this? Because the premises in such circumstances represent something richer in information and more controversial than the conclusion (the informational riches of which we already acquired as children). We do not even try to convince somebody that ice floats on water. We do not even want to prove it, as all of us have believed it from the beginning. We try to explain this conviction: more precisely, we want it to result from the premises of a deductively valid argument as a conclusion to it. According to Dummett, circularity in this case should not represent anything interesting or important. On the other hand, the existence of concrete circumstances under which circularity does not represent anything interesting or important significantly disturbs the universal claim of the logical reason behind the fourth point. Because of these circumstances, its unrestricted impact seems to be rather out of the picture.

In turn, the pragmatic reason for the fourth point is challenged by Crispin Wright in "Intuition, Entitlement and the Epistemology of Logical Laws" (Wright 2004b). Based to a significant degree on the thought stimuli contained in the first three points, Wright forms the basis for a plausible theory of logical knowledge. This means that, in direct opposition to Wittgenstein's fourth point, he tries to present a coherent and rational explanation of what serves as a support for our disposition for following basic logical norms and for our knowledge of the validity of derived logical norms.

From Wright's perspective, we should not forget that the pragmatic reason for the fourth point is based entirely on "traditional ideas" about what should constitute a rational explanation or a theory (that is, what the determinate and the indeterminate should be like). According to Wright, we should not let the tradition restrict us to such an extent. Many self-evident ideas about what should constitute a theory (what the determinate should be like) have gradually turned out to be "naïve". In other words, Wright agrees with Wittgenstein that the "classical" theory of the knowledge of the validation of logical norms will most probably remain undiscovered. In contrast to Wittgenstein, he does not perceive it as an unresolvable paradox. In fact, this does not mean that we cannot discover some other form of coherent rational explanation for the abovementioned validation (other than a classical one), and that, on this basis, we will understand to a sufficient extent the entitlements and reasons which our knowledge of logical norms is based upon. The failure of traditional theoretical presuppositions in itself does not mean that a form of theory other than that which we expected cannot exist. And this is a mistake, because it actually can exist (Dvořák 2016, 232–241). Wright's realization introduces another disturbance to the universal claims of the pragmatic reason for the fourth point. Because of this, its unrestricted impact also seems to be rather out of the picture.

Upon the pretext of these reservations, I will attempt to make my own evaluation of the fourth point. I agree with Dummett that circularity should not be viewed as such a big problem. There are contexts in which circularity is not controversial at all (for example, during an attempt at an explanation of a conviction which all the participants have agreed upon). I would add that the existence of a context in which circularity is not controversial undoubtedly undermines the status of the "logical reason" for the fourth point, which Wittgenstein sees as unquestionable. I also agree with Wright that the failure of traditional ideas about what should constitute a rational explanation does not mean that there is no possibility of some completely different form of rational explanation aside from that which we expect. On the contrary, such a new (unconventional and unexpected) form can easily come into existence. Again, it is true that this recognition seriously undermines the status of the "pragmatic reason" for the fourth point used by Wittgenstein as a solid foundation.

Both illustrations represent a problem for the stance in question. Upon their basis, it seems that it would be irresponsible and unjustified to give up the project of the theory of the origin of criteria we usually use when making decisions too quickly. Even if we had to accept some unorthodox ideas and "unthinkable" solutions for that purpose. I am convinced that we should not feel any unnecessary inhibitions in that regard (Schneider 2014, 2–4). It is specifically from this vitalist and optimistic perspective that the fourth point seems rushed and overly pessimistic.

Conclusion

I will now try to summarize the content of this investigation. I will first summarize Wittgenstein's attitude and then my objections to it. According to Wittgenstein, if we realize what the content of the first to fourth points means, we will surely cease to feel the necessity of searching for answers to questions such as "Where did the activity of rule-following (or the refusal of rules) that we see all around us come from?" Such matters should not interest us as theoretical questions. In connection to questions of this kind, Wittgenstein believes that we do not need to have a theory or an explanation at our disposal but rather something categorically different. We need to correctly see what is happening around us. We need to look at the questions we pose from the right perspective. Why is this? Because thanks to such an approach, we should also be able to better understand our own relationship with rules, regulations, commandments, and so on. Consequently, we could better understand our own place in the world. And ourselves as well. Herein there lies, I dare say, the primary motivation of Wittgenstein's notes on normativity. I therefore reach the conclusion that, in his reflections from the later period, he tries to defend a specific anti-theoretical approach to the question of the origin of the criteria that normally govern our decision-making.

Nonetheless, it is important in this regard to stress that he does not promote animosity towards the construction of theoretical systems or an apathetic approach towards philosophical disputes. On the contrary, we see here the result of a specific realization that he reached through a critical analysis of normativity itself. It is therefore not a defence of scepticism nor relativism but rather a recognition of something like the boundaries of the theoretical. More precisely, it is the result of pushing these boundaries.

I do not doubt the existence of such boundaries. At the same time, however, I am convinced that in Wittgenstein we encounter an overly one-sided and overly pessimistic attitude to this phenomenon. In fact, a completely different approach to boundaries in general and the boundaries of the theoretical is possible. How can this be done? By seeing them as something that is persistent but not permanent. This means that we can see them as something that we are given the task of breaking, working our way around, or jumping over, and not as something to subjugate ourselves to (even though we may sometimes "crack our head" in the process). Opting for such a vitalist and optimistic approach, as well as seeing the boundaries of the theoretical as something which is ultimately supposed to open the way to new dimensions and worlds – and not conceal them from us – is quite possible (Kvasz 2015, 169–194). That is the crucial point here. I therefore reach the conclusion that had Wittgenstein been aware of this truth, he would have had to abandon his anti-theoretical stance.

Acknowledgement
This chapter was written at the Department of Philosophy and Applied Philosophy, Faculty of Arts, University of Ss. Cyril and Methodius in Trnava, as a part of VEGA project 1/0563/18.

Bibliography

Ayer, Alfred Jules. 1966. "Can There Be a Private Language?" In *Wittgenstein: The Philosophical Investigations*, ed. by George Pitcher, 251–266. New York: Doubleday and Company.

Boghossian, Paul. 2000. "Knowledge of Logic." In *New Essays on the A Priori*, ed. by Paul Boghossian and Christopher Peacocke, 229–254. Oxford: Clarendon Press.

Coliva, Annalisa. 2015. *Extended Rationality: A Hinge Epistemology*. New York: Palgrave Macmillan.

Davidson, Donald. 2006. "On the Very Idea of a Conceptual Scheme." In *The Essential Davidson*, Donald Davidson, 196–208. Oxford: Clarendon Press.

Descartes, René. 2005. "Meditations on First Philosophy." In *The Philosophical Writings of Descartes, Vol. II.*, trans. by John Cottingham. Cambridge: Cambridge University Press.

Dummett, Michael. 1978. "The Justification of Deduction." In *Truth and Other Enigmas*, Michael Dummett, 290–318. London: Duckworth.

Dvořák, Petr. 2016. *Logika onticky neurčitých domén: Jsou logické pravdy náhodilé?* Prague: Togga.

Horwich, Paul. 2012. *Wittgenstein´s Metaphilosophy*. Oxford: Clarendon Press.

Kripke, Saul. 2002. *Wittgenstein on Rules and Private Languages*. Oxford: Basil Blackwell.

Kvasz, Ladislav. 2015. *Inštrumentálny realizmus*. Pilsen: Pavel Mervart.

Maddy, Penelope. 2014. *The Logical Must, Wittgenstein on Logic*. Oxford: Oxford University Press.

Moyal-Sharrock, Daniele. 2017. "Wittgenstein on Knowledge and Certainty." In *A Companion to Wittgenstein*, ed. by Hans-Johann Glock and John Hyman, 547–562. Chichester: Wiley Blackwell.

Nietzsche, Friedrich. 1968. "The Birth of Tragedy." In *Basic Writings of Nietzsche*, trans. by Walter Kaufmann and Reginald J. Hollingdale. New York: Modern Library.

Nietzsche, Friedrich. 1997. *On the Genealogy of Morality*. Trans. by Maudemarie Clark and Alan J. Swensen. Indianapolis: Hackett Publishing Company.

Putnam, Hillary. 1981. *Reason, Truth and History*. Cambridge: Cambridge University Press.

Rheese, Rush. 1966. "Can There Be a Private Language?" In *Wittgenstein: The Philosophical Investigations*, ed. by George Pitcher, 267–285. New York: Doubleday and Company.

Schneider, Hans Julius. 2014. *Wittgenstein's Later Theory of Meaning: Imagination and Calculation*. Oxford: Wiley Blackwell.

Schönbaumsfeld, Genia. 2016. *The Illusion of Doubt*. Oxford: Clarendon Press.

Wittgenstein, Ludwig. 1958. *The Blue and Brown Books: Preliminary Studies for the Philosophical Investigations*. Oxford: Blackwell Publishers.

Wittgenstein, Ludwig. 1971. *Tractatus Logico-Philosophicus*. Trans. by David F. Pears and Brian F. McGuinness. London: Routledge and Kegan Paul.

Wittgenstein, Ludwig. 1972. *On Certainty*. Trans. by Denis Paul and Gertrude Anscombe. New York: Harper and Row Publishers.

Wittgenstein, Ludwig. 1999. *Philosophical Investigations*. Trans. by Gertrude Anscombe. Oxford: Blackwell Publishers.

Wright, Crispin. 2004a. "Wittgensteinian Certainties." In *Wittgenstein and Scepticism*, ed. by Denis McManus, 22–55. London: Routledge.

Wright, Crispin. 2004b. "Intuition, Entitlement and the Epistemology of Logical Laws." *Dialectica* 58, 1: 155–175. DOI: https://doi.org/10.1111/j.1746-8361.2004.tb00295.x.

Chapter 4

The Application and Understanding of Default Autonomy in Ethically Dilemmatic Cases Presented by Czech Medical Doctors: An Empirical Study

Martin Zielina, Jaromír Škoda, Adam Doležal, Barbora Beňová, Kateřina Ivanová, and Adéla Lemrová

Abstract: Beauchamp and Childress proposed a theory of autonomous action that is currently considered the default concept of autonomy in bioethics. According to their theory, the following conditions need to be met for action to be truly autonomous: (i) the agent acts intentionally, (ii) with understanding, and (iii) without any controlling influences that determine their action. It has been established that the concept of default autonomy (DA) should be the basis of the doctor–patient relationship. The presented empirical study aims to assess the ability of Czech medical doctors to meet the concept of DA in ethically dilemmatic cases. Fifty-two out of sixty-nine cases were evaluated utilizing an interpretative phenomenological analysis and the Four Boxes model to determine whether (i) all cases met the criteria of default autonomy; (ii) if not, which criteria were omitted; and (iii) what was the most commonly omitted criterion of DA. Then we classified the cases into three categories based on the number of criteria fulfilled. We found that only 21% of cases met all three criteria of DA. The criteria omitted most frequently included intentionality (35%), understanding (26%), and voluntariness (25%). Twenty-one percent of cases were classified as a "white zone", meaning that all criteria of DA were met; sixty percent of cases were classified as a "black zone", where at least one criterion was not met; and nineteen percent of cases were classified as a "grey zone", where we could not determine whether all criteria had been met or not.

Keywords: Default autonomy, doctor–patient relationship, informed consent, competence, medical decision-making.

Introduction

For a long time, a paternalistic relationship between a doctor and a patient has been the norm. Only in the second half of the twentieth century did a novel patient-centred approach emerge; this was an approach that was centred on a respect for the patient's autonomy. Even though the principle of respect for a patient's autonomy had been present in the legal and bioethical context in earlier times,[50] it became dominant only after the Second World War and the accompanied technological advances that allowed for choice between multiple treatment options. After the fatal failures of medical doctors during the Second World War, the focus

[50] The two major distinct interpretations of a patient's autonomy in medical ethical literature from a historical perspective are presented by Jay Katz (2002) and Martin S. Pernick (1982).

on respect for persons gained an important momentum, initially in the context of clinical research[51] and eventually in clinical practice as well. As a right to self-determination, autonomy has become the dominant principle of modern bioethics. Beauchamp and Childress, who were the founding fathers of ethical principlism, defined the current (and most frequently used) understanding of the principle of autonomy; as a result, their concept of autonomy became known as the "default autonomy" (DA).[52] According to the concept of DA, a person's decision can only be considered autonomous if made *intentionally,* with a *substantial level of understanding*, and *without significant external controlling influences* (Beauchamp and Childress [1979] 2009, 101n).

In the presented chapter, we chose the above-described minimal concept of autonomy as the baseline for our analysis. However, instead of examining cases from a "top-down" perspective, we decided to perform a bottom-up analysis of cases to determine whether they met the three criteria of DA by means of an interpretative phenomenological analysis (IPA). The concept of DA refers to the autonomy of *action* (autonomy of *choice*) rather than to the autonomy of a person. Under certain conditions, an autonomous person can act in a non-autonomous way, whereas a non-autonomous person can act in an autonomous way in some situations.

The definition of DA states that an act is performed in an autonomous way only if the agent acts intentionally, with understanding, and without controlling influences (Faden and Beauchamp 1986, 238). Each criterion represents a necessary condition of an autonomous action, and collectively these criteria are sufficient to meet the definition of DA.

A person can act intentionally or non-intentionally; intention is either present or absent. If an action is nonintentional, it is also necessarily non-autonomous. However, a person can possess a certain level of understanding, and a greater level of understanding leads to a more autonomous action. The same applies to the extent of an external controlling influence; the weaker the influence of external circumstances, the more autonomously one may act.

The extent of understanding spans from a full understanding to a complete lack of it. In order to ascertain whether a person acts in an autonomous way, *a substantial level of understanding* needs to be achieved. This *substantial level of*

[51] See international ethical codes such as the Nuremberg Code (1947) and the Declaration of Helsinki (1964). The Nuremberg Code states in its first sentence that "the voluntary consent of the human subject is absolutely necessary".

[52] The concept of default autonomy is developed most thoroughly in the works of Ruth Faden and Tom Beauchamp (1986).

understanding remains difficult to define. Many authors acknowledge the presence of a *grey zone*; they therefore recommend a *threshold of understanding* to be set that would indicate a substantial level of understanding to meet the "understanding" criterion of autonomy. Analogically, the degree of external controlling influence may vary. The degree of control determines whether an action is autonomous or not. The concept of a *grey zone* applies here too. Faden and Beauchamp opine, that *coercive situations* – e.g. illness, pain, medication, addiction, and economic and social pressures – do not restrict a patient's autonomy. Even though such coercive situations may impose psychological limits on a patient's freedom to decide, they do not restrict the patient's ability to act in an autonomous way. Since many authors disagree with Beauchamp's and Faden's stance towards the effect of coercive situations on a patient's autonomy, we decided to consider coercive situations as relevant external controlling influences (Fig. 1).

Figure 1: Degrees of autonomy of intentional actions

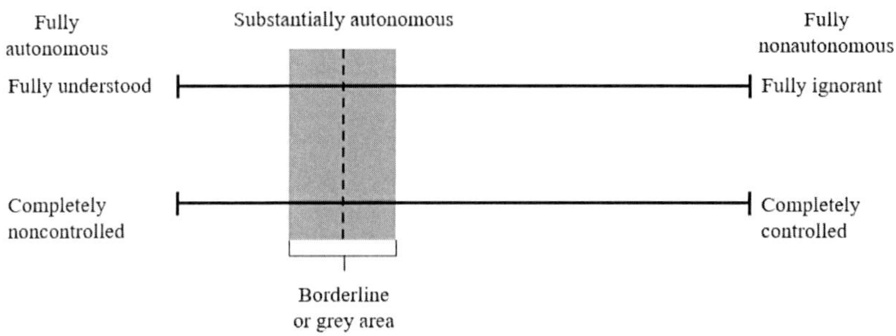

From: Faden and Beauchamp (1986, 239).

Competence is a concept that is separate from the abovementioned criteria of DA, and Beauchamp, Childress, and Faden do not clarify the relationship between autonomy and competence in their works (Beauchamp and Childress [1979] 2009, 113n; Faden and Beauchamp 1986, 287n). However, we considered it vital to include competence in our analyses.

Competence is a feature of persons whose autonomous decisions must be respected. Persons who lack competence need to be limited in their decision-making on the account of the principle of beneficence. Therefore, competence judgment functions as a *gatekeeper,* limiting the spectrum of those allowed to make decisions for themselves (*the gatekeeping function of competence judgments*). No autonomous action can take place without the competence of an acting agent;

therefore, for our purposes, competence can be understood as a sufficient level of abilities necessary for autonomous action. Competence represents a potentiality for and a necessary condition of the realization of autonomy; autonomous action represents an actual realization of competence.

A. Competence

A person is competent to perform a certain action only if they have all the necessary capacities to perform it; for the purposes of our study, this meant the capacities to make decisions about a proposed course of treatment. Competence is context-dependent and depends on things like the stage of a patient's disease, external surroundings, and a patient's current physical and mental condition.

In general, the necessary capacities that determine a person's competence comprise:
(a) the ability to receive information and to understand it
(b) cognitive abilities necessary to consider alternative courses of action (in a medical context)
(c) decision-making capacities that include the ability to make a decision, to commit to a decision despite having doubts, and to clearly express it.

In medical practice, a patient may lack competence under specific conditions (Wear 1993, 49; Berg et al. 2001, 99), such as if they are a minor, if they suffer from a neuropsychiatric condition that limits cognitive functions (e.g. a type of dementia), or if their cognitive abilities are affected by medication. A patient can have their competence limited by a court order or they may refuse recommended treatment for apparently irrational reasons (Eth 1985). However, even competent patients are entitled to make irrational decisions, and an irrational decision may represent an autonomous expression of a competent person's free will despite medical personnel's doubts.[53]

Given the ethical and legal complexities in the assessment of competence, in practice a person is assumed to have competence unless proven otherwise (*presumption of competence*) (*Mental Capacity Act* 2005).

B. Intentionality and intentional action

Intention[54] represents an internal state of mind and the content of one's conscious mind, whereas intentional action represents an expression of the intention in the

[53] Some authors dispute whether an irrational decision made by an otherwise competent patient should be respected. See Lesser (1983, 144).

[54] The terms "intention" and "intentionality" are often associated with the philosophy of mind in which the philosophers discuss the issue of how our mental states reflect the objects in the world. This specific

external world through the action of an individual. Intention comprises volition, planning, the anticipation of an action's consequences, and the purposeful execution of an intention.[55] Indirect intention means that a person intends a consequence (obliquely) when that consequence is a virtually certain consequence of their action, and they knew it to be a virtually certain consequence.[56]

In medical practice, intentionality comes under scrutiny in cases of inadequate informed consent, i.e., whether a patient would have decided otherwise (usually against a specific procedure) providing they had had an adequate level of information.

C. A patient's understanding

As previously mentioned, understanding is deeply intertwined with intentionality. No one can make a decision intentionally without having an adequate level of understanding and consequently acting in an autonomous way. In the practice of medical ethics and medical law, the affirmation of intentionality is usually derived from an adequate level of understanding and the absence of external controlling influences (Faden and Beauchamp 1986, 299).

Also, we need to emphasize the distinction between understanding and competence. The capacity to understand the disclosed information represents a key condition of competence;[57] therefore, we can only evaluate a competent person's (level of) understanding. A person may understand the information disclosed to a certain extent. A problem arises if the information was disclosed in an incomprehensible way, such as by using specific medical terminology. Initially, *the disclosure of information* was scrutinized in the medical-legal setting rather

---

meaning of the terms "intention" and "intentionality" was broadly analysed in the philosophy of Franz Brentano, Edmund Husserl, and later on in analytical philosophy, such as in the works of John Searle and Daniel Dennett. See Brentano (1973); Searle (1983) and Dennett (1987). Here, however, we use the original meaning of "intention" related to the intention of an acting agent. See Nadelhoffer (2008, 2).

[55] There are two distinct types of intentions: an *end-directed intention* and a *simple intention*. Whereas an *end-directed* intention aims to achieve a state of affairs ϕ by means of act A, the aim of a *simple* intention is merely act A itself. See Audi (1973, 387).

[56] Ruth Faden and Tom Beauchamp (1986) present the example of a person whose indicator is not functioning and they want to signal a change of direction by using their arm; however, it is raining outside. The person does not want and does not wish to get their arm wet, but they do want to signal a change in direction. Therefore, they have an indirect intention (or are aware of the fact) that their arm will get wet. The distinction between direct and indirect intention is especially important in the context of Anglo-American and European law.

[57] A different perspective is presented by Grisso and Appelbaum (1998, 37n). The authors argue that competence is context-dependent and therefore falls under the concept of understanding. The level of understanding can increase if the information is repeated or re-explained in a more comprehensible way, or if the external surroundings are adapted accordingly. In this case, the concept of understanding does not represent an independent entity but rather falls under the concept of competence.

than the patient's *actual understanding* of the information provided. The situation has changed, and today the emphasis is placed on the fact whether a patient *understood* the information rather than whether the information was disclosed.[58]

D. Significant external controlling influences

In order to meet the criteria of DA, an action needs to be performed voluntarily. According to current theories, an action is voluntary if (and only if) it is not a result of coercion or a lack of knowledge (Hyman 2013). Faden and Beauchamp (1986, 259) define the following external controlling influences: (a) coercion, (b) manipulation, and (c) persuasion. There exists a grey zone between these three types of influences.[59] While coercion always results in a non-autonomous action, and persuasion is acceptable, the effect of manipulation is always difficult to evaluate regarding its effect on DA (Fig. 2).

Figure 2: The continuum of influences from controlling to noncontrolling ones

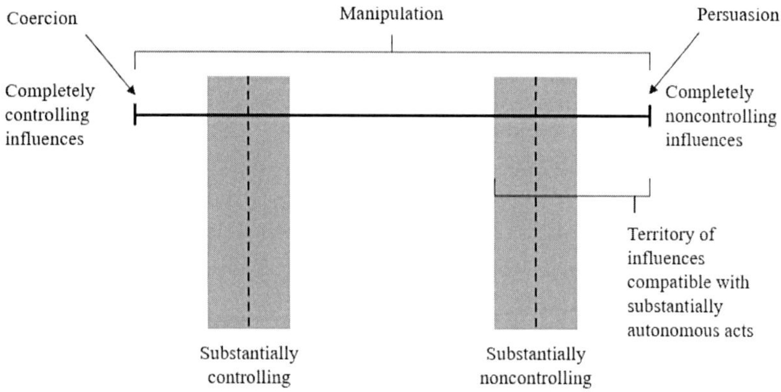

From: Faden and Beauchamp (1986, 259).

For the purposes of the presented study, the following criteria constitute the concept of DA: (a) competence, (b) intentionality, (c) a sufficient level of understanding, and (d) absence of external controlling influences. Analysing ethically

---

[58] There is an ongoing debate about whether a one-sided *disclosure* of information is sufficient or whether it is necessary to initiate a dialogue in which the medical doctor provides the information on medical indications and the patient informs the doctor about their values, preferences, life plans, and so on. In medical ethics, the second model of *shared decision-making* is traditionally preferred. See Charles et al. (1999).

[59] Since the effect of manipulation is difficult to evaluate, the law distinguishes only two types of influence (justified and non-justified). This is in contrast to the three types of influence presented by Faden and Beauchamp.

dilemmatic cases in light of these criteria, we aimed to determine (i) whether default autonomy was maintained in the analysed ethically dilemmatic cases, (ii) which criteria of DA were not met in cases where DA was not maintained, and (iii) which of the three criteria of DA was omitted the most frequently.

METHODS

Data collection and analysis

Ethically dilemmatic cases were collected from February to July 2019 using a questionnaire with open questions administered to medical doctors during lectures on medical ethics that took place during their residency training. The participants provided their answers voluntarily after a thorough explanation of the aims and methods of the survey and under the condition of complete anonymity for the participants themselves and those in the described cases.

Questionnaire design

The questionnaire was developed using the Four Boxes model (Jonsen et al. 2015). This model originated in casuistry and, along with principlism, these two represent the most common approaches to decision-making in bioethics, especially in North America and Western Europe (Table 1). Based on this model, the questionnaire included open questions related to ethical aspects of the case itself; the reported patient's past medical, social, and family history; and the level of involvement of the reporting physician in the case (directly involved as the decision-making agent, a direct witness, or an indirect witness not involved in the case). The original Four Boxes model was modified for the purposes of the presented study with the aim to take into consideration local specificities (i.e. the context of a medical ethics course for junior doctors in residency training in the Czech Republic) and to capture the complexity of the dilemmatic case.

Table 1: The Four Boxes model and principlism

Four principles (Beauchamp and Childress [1979] 2009)	*Four Boxes* (Jonsen et al. 2015)	
Respect for Autonomy Beneficence	Medical Indications (beneficence and nonmaleficence)	Preferences of Patients (respect for autonomy)
Nonmaleficence Justice	Quality of Life (beneficence, nonmaleficence, respect for autonomy)	Contextual Features (justice)

From: Ross (2015, 270).

IPA as a tool for case analysis

IPA originates from three major sources: phenomenology, hermeneutics, and the idiographic approach (Řiháček et al. 2013). Phenomenological sources emphasize the effort to describe phenomena without interpreting them (Pringle et al. 2011). Nevertheless, researchers' preconceptions are accepted as long as they enable them to formulate the meaning of the participant's experience (Fade 2004). Hermeneutics provides a "double hermeneutics" approach that takes into consideration the fact that the participant (a physician) tries to understand their experience (an ethically dilemmatic case) as well as the fact that the researcher aims to understand the process of how the participant came to this experience. Finally, the idiographic approach emphasizes the uniqueness of an individual (the participant) and their specific situation. Individual physicians vary in their pursued medical specializations, their patients and their medical needs, treatment options and courses, and their unique life experience.

IPA poses questions on how individuals or groups experience certain situations and what meanings they ascribe to them (Smith and Osborn 2003). In such cases, research questions include expressions such as "experience" and "lived experience". However, IPA also utilizes secondary research questions that verify the accordance between a personal experience and a theory (Smith et al. 2009). It was therefore our aim to analyse the experience of Czech medical doctors with the application of patients' DA in situations that they (medical doctors) themselves consider ethically dilemmatic.

We focused on the analysis of the experience of Czech medical doctors with ethically dilemmatic cases, and we evaluated them in relation to the concept of default autonomy based on three criteria of DA: (1) a patient's action was intentional and the patient was competent; (2) a patient's action was based on their understanding as evidenced by a sufficient level of information provided, including the information about other available treatment options; and (3) a patient's action was free from significant external controlling influences (Table 2).

Table 2: The criteria of DA

1. Intentionality		2. Understanding			3. Controlling influences
Compe-tence	Intentional action	Sufficient amount of information	Infor-mation provided	Disclosure of all available treatment options	Controlling influences
(1) yes (0) no	(1) yes (0) no	(1) yes (2) rather yes (3) rather no (4) no	(1) yes (0) no	(1) yes (0) no	(1) yes (0) no

An analysis of competence, intentionality, and intentional action

For an analysis of competence, intentionality, and intentional action, we assessed whether a patient was competent and whether they expressed their will regarding a future course of treatment or whether they expressed disagreement with a proposed course of treatment. Competence was presumed in all patients unless reported otherwise (e.g. a patient in a coma, or a patient with the diagnosis of dementia who did not respond adequately to questions). In contrast, intentional action was not presumed unless explicitly stated using expressions such as "the patient decided that...", "intended to...", "wanted to...", "did not wish to...", and "refused".

An analysis of the patient's understanding

In the analysis of the patient's understanding, we presumed that information was not provided unless explicitly stated that the physician had provided the information on the patient's course of treatment. If the participant stated in the questionnaire that multiple treatment options were available and disclosed to the patient, the case was scored in the affirmative. For the assessment of the level of patient's understanding, we employed a qualitative approach to assess whether the information was provided to a sufficient extent and in an adequate way (yes, rather yes) or not (no, rather no).

An analysis of controlling influences

In Faden's and Beauchamp's view, DA can be limited by significant external controlling influences only when exerted by other persons. In our cases, we encountered situations in which the participants mentioned external influences, e.g. information was not disclosed to a patient at the patient's family's request.

A complex case data analysis

Data analysis was performed in three consecutive phases. During the first stage, the reviewers (MZ, JS, and AD) evaluated sixty-nine cases independently using a self-designed scoring tool that combined a qualitative (IPA) and a quantitative approach (Table 2). This tool enabled us to describe cases in sufficient detail (a qualitative approach) and to quantitatively analyse the criteria of DA for all cases. In the second stage, the same reviewers analysed the level of agreement for specific categories in the case description that had been achieved in the first-stage analyses. The overall agreement reached 50.6%; this was lower for certain categories (e.g. "disclosure of all available treatment options" – level of agreement 30%). The lack of agreement was caused by equivocal formulation of certain categories (e.g. "disclosure of all available treatment options" and "the amount of information provided to the patient") and by different interpretations of ethical vs. legal meanings of other categories (e.g. "competence" and "intentional action"). The less reliable categories were re-scored, and the scores in the reliable categories were retained.

During the third stage, data triangulation was performed. This consisted of the final unification of scores in respective case categories from the first and second stages. The entire process led to an increase in coding reliability, and the agreement reached the level of "complete agreement" in 92% of cases and "partial agreement" in 8% of cases; no case was classified as "lack of agreement". For specific categories, the minimum level of agreement reached 83.6% ("sufficient amount of information provided to the patient"), and the maximum level of agreement reached 98.1% ("patient's competence").

Afterwards, we categorized the cases into three groups. Cases where all three criteria of DA were met were included in the "white zone" group. Cases where at least one criterion of DA was omitted were included in the "black zone" group. Cases where we were unable to determine whether all or some criteria were met or omitted were included in the "grey zone" group.

RESULTS

A characterization of the studied cohort

Altogether, we obtained sixty-nine completed questionnaires, out of which seventeen cases were eventually excluded. The reasons for excluding specific cases could be summarized as follows:

1. The response in the questionnaire did not describe an actual case (n = 10), e.g. the participants described a general ethical dilemma, or presented some features

of a case, but did so without sufficient detail and leaned towards a general description of an ethical dilemma.
2. The case included two agents, and it was unclear whose DA was being questioned (e.g. cases of maternal-foetal conflicts, n = 4).
3. The reporting physician was not physically present or directly involved in the situation described (n = 3).

The average age of participating medical doctors was thirty-one years (a median age of thirty years), and 55% of participants were female. The average age of patients reported in cases was sixty-two years (a median age of sixty-nine years). In eighteen cases, age was not disclosed. Forty percent of reported patients were female and 37.7% were male; in the remaining cases, the sex was not disclosed.

B. The criteria of DA

The most frequently omitted criterion of DA was intentionality, which was absent in 35% of cases. Absent or diminished competence was reported in 19% of cases and the patient was not adequately informed in 26% of cases. No information was provided to the patient in 15% of cases, and the patient was subject to external controlling influences in 25% of cases (Table 3).

Table 3: The fulfilment of criteria of DA

1. Intentionality		2. Understanding			3. Controlling influences
Competence	Intentional action	Sufficient amount of information	Information provided	Disclosure of all available treatment options	Controlling influences
%	%	%	%	%	%
(1) 73	(1) 54	(1) 33	(1) 52	(1) 44	(1) 25
(0) 19	(0) 35	(2) 12	(0) 15	(0) 48	(0) 69
		(3) 13			
		(4) 13			
N/A 8	N/A 11	N/A 29	N/A 33	N/A 8	N/A 6

Cases were categorized into three groups (Fig. 3) based on whether the criteria of DA were met. All DA criteria were met in 21% of cases (white zone), at least one criterion was omitted in 60% of cases (black zone), and in 19% of cases we were unable to determine whether all or any of the criteria were met (grey zone).

Figure 3: Three zones of cases in relation to the fulfilment of DA criteria (n = 52)

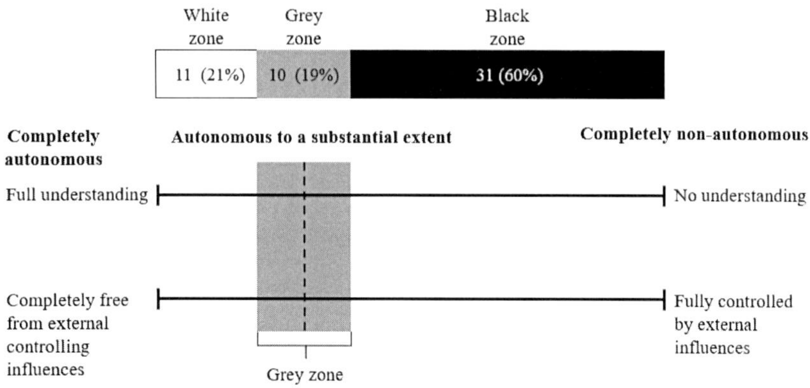

Adapted from: Faden and Beauchamp (1986, 239) (modified by the present authors).

The white zone cases

Below we present a transcript of a case that meets all the criteria of DA. The patient was presumed to be competent (no information on lack of competence was reported). She expressed her will voluntarily without any external controlling influences; she refused the proposed medical procedures. She was adequately informed about all available treatment options (aortic valve surgery or a transcatheter replacement of the aortic valve) and their potential complications and consequences.

Case 1: The patient is a ninety-one-year-old female. Her past medical history includes surgery for uterine myomas, and she is being treated for hypertension. No known allergies. She lives in a detached house with her daughters who take care of her. The patient suffers from orthopnoea (a resting shortness of breath) without a known cause. In addition, she started having feelings of dizziness, episodes of falls, and she stopped walking. She was diagnosed with aortic stenosis. Her medical issues could be resolved either by aortic valve surgery or by aortic valve implantation (TAVI – transcatheter aortic valve implantation). The cardiothoracic surgery consultant indicates TAVI, but this decision could be challenged given the patient's age even though the patient has no significant comorbidities or dementia. The consultants in critical care and internal medicine also consider the patient eligible for surgery. However, there is a significant risk of the patient becoming ventilator-dependent even after a simple surgery. In contrast to her family, the patient does not want the surgery. The patient is informed and she refuses both the surgery and the TAVI.

The black zone cases

Here we present cases in which one or more criteria of DA were not met. In Case 2, the patient was not competent due to loss of consciousness and he did not express his medical preferences in advance. In Case 3, the patient changed her preferences but her wish was not respected due to external controlling influences (the patient's relatives). In Case 4, the patient was not sufficiently and adequately informed about the proposed surgical procedure.

Case 2: The patient is a sixty-nine-year-old male with a supportive family environment. He was acutely admitted for a ruptured aneurysm that resulted in massive intracranial bleeding with a fatal prognosis; surgery was not indicated. The physician in charge indicates no further invasive procedures, the treatment of pain, and the dignified death of the patient in a coma. However, the patient's family refuses to accept the diagnosis and the suggested treatment plan; they are aggressive and threaten the medical team despite repeated conversations that aim to elucidate the patient's condition. The family insists on "saving" the patient's life at all costs. Under the influence of repeated threats and "important names", the consultant in charge decides to perform intubation and to initiate mechanical ventilation. The patient's dying is prolonged from two to three days to almost three months in the intensive care unit; the patient is in a coma, suffers from infections, bedsores, tubes, catheters, urine, stool, secretions, and so on with financial ramifications. After three months of the patient's "suffering", the family complains about "bad care", "bad doctors", "bad nurses", and "bad health care".

Case 3: The patient is a thirty-four-year-old female: university-educated, a schoolteacher, and primipara. The patient arrives with labour plans in the initial phase of labour. The foetal monitor shows normal findings without a membrane rupture. The aim was for the patient to give birth with as few medical interventions as possible. The patient's family was informed that if the medical conditions warranted an intervention, or if the patient herself changed her preferences, the labour plan might not be followed since it is the first time she is giving birth and she will see what would be necessary (e.g. pain medication, positioning, or a foetal monitor). The labour plan stated that the medical team should only communicate with the patient's partner and not disturb the patient. The labour plan also stated that the partner can make decisions about suggested medical procedures since the patient would be (according to the family) unable to decide for herself.

During the course of labour, non-pharmacological means of pain control failed, the labour did not proceed, and the patient began to request epidural anaesthesia for severe pain in strong contractions. Her partner, however, insisted that she did not need anaesthesia and that the pain was only in her mind. It took several hours of conversations and explanations until the partner allowed epidural anaesthesia. The ethical aspect of this case involved the patient's quality of life and whether to administer epidural anaesthesia contrary to the labour plan (validated by a notary) and the partner's will.

Case 4: The patient is a thirty-six-year-old female: university-educated and a schoolteacher. The patient's mother died at the age of forty-five of colorectal carcinoma. The patient underwent a colonoscopy at her own request because she was afraid that she might have the same tumour as her mother did. Everyone laughed at her for having a colonoscopy, saying that she was young and definitely did not have a tumour. The colonoscopy was performed at a department of internal medicine in a local hospital, and in fact it disclosed the presence of a colorectal carcinoma. The local consultant in internal medicine simply printed out the colonoscopy findings and referred the patient to a surgical department. He did not inform the patient about the findings or explain a further course of treatment. The surgery consisted of a partial resection of the affected colon (hemicolectomy). The doctor at the surgical department who was admitting the patient was a resident in internal medicine on a rotation at the surgical department, and he did not feel competent enough to explain to the patient all the details regarding the surgery. Therefore, he asked a colleague nearby, who was a consultant in surgery, to provide an explanation. The consultant ironically stated: "Well, we remove half of your bowel and the nodes, and maybe you will have a stoma," and he left. The woman collapsed and started crying. The resident tried to calm her down and started explaining everything again from the perspective of internal medicine.

The grey zone cases

In "grey zone" cases, we were unable to determine whether one or more of the DA criteria were omitted, even after data triangulation in the third stage. It might have been caused by the inherent limitations of the questionnaire or by insufficient case description. In the following case, we might presume the patient's competence; however, the patient does not desire further medical examinations. The patient's intentional request prevents the medical team from providing further information, and we cannot therefore ascertain whether all DA criteria were met.

Case 5: The patient is an eighty-year-old woman who is lucid. She suffers from chronic and terminal complaints. About ten years ago, she had breast cancer; currently she has multiple metastases in her bones and abdominal cavity as well as pathological fractures of both humeri, with one of them being dislocated. At the orthopaedics ward, she receives pain treatment and a surgical fixation of the dislocated bone. The oncologists suggest further examinations, including a gynaecological exam. The patient probably felt that she had an incurable disease, and she did not wish to know any specific information and did not want further examinations (oncological or gynaecological).

DISCUSSION

In the presented study, we focused on the concept of DA in ethically dilemmatic cases collected among Czech medical doctors during lectures in medical ethics that represent an obligatory part of their residency training. We are aware of the theoretical, methodological, and interpretational limitations of this study.

The theoretical limitations include the limitations of the concept of autonomy itself. Experts generally agree that respect for the patient's autonomy needs to be upheld; however, it remains unclear what that means in specific cases. Informed consent, although instrumental in supporting the patient's autonomy, may not be sufficient (Berg et al. 2001). Critics also rightly argue that bioethical principles, including respect for the patient's autonomy, represent a mere "anthology" or sum of principles without a hierarchy, and it is unclear which one takes precedence over the other in conflicting cases (the "anthology syndrome") (Gert et al. 1997).

The methodological limitations stem mainly from a variety of empirical approaches focused on the study of patients' decision-making (Say et al. 2006). These studies differ in their methodologies, participant selection, and results. The authors of the presented study are not aware of an empirical study focused on clinical decision-making in the Czech context. In addition, the empirical approach has not yet been clearly established in bioethics even though empirical studies are on the rise and their proportion has increased from 5.4% in 1990 to 15.3% in 2003 in nine major bioethical journals (Borry et al. 2006). Wangmo and Provoost (2017) argue that empirical research has its place in bioethics; in their study, 193 of 200 bioethicists from twelve European countries agreed that empirical research is appropriate for a contextual description of an ethical problem. However, disputes remain over the normative value of results from empirical studies in bioethics, and they warrant an extensive debate in the future (Davies et al. 2015). In our study, we have shown that the methodology combining casuistry and principlism represents a viable approach for the collection and analysis of ethically dilemmatic cases and could be used for further and more extensive studies in the future. The abovementioned limitations could also influence how we interpret patients' perceptions and how we assess their willingness to participate in the decision-making process. Strull et al. (1984) found a stark contrast between patients' willingness to make decisions and their doctors' perception thereof. The patients preferred their doctors to decide to a much greater extent than what their doctors thought.

Conclusions

Every doctor–patient relationship is unique. However, even in unique situations, doctors and patients should strive to uphold a patient's DA. In the presented study, we found that the majority of reported cases did not fulfil all of the criteria of DA and that in some cases more than one criterion was omitted; intentionality was the most frequently omitted criterion. Multiple studies conclude that patients of a younger age, with higher levels of education, and preferentially women tend to be more active in their medical decision-making. Arora and McHorney (2000) reported on a population of 2197 patients and asserted that patients with less serious medical conditions (e.g. mild hypertension) tend to be more involved in medical decision-making than those with more serious ones (e.g. advanced diabetes and serious heart conditions). In our study, we were unable to consider all of these contextual features; however, we aim to focus on them in more detail in the future.

The categorization of cases in white, black, and grey zones should not be understood in a normative sense and should not lead to accusations of unethical conduct. A lack of DA in certain cases results from the patient's condition itself (e.g. a patient in a coma without an advance directive) or from a complex of intertwined contextual features. The proposed categorization aimed to shed more light on the complexity of medical decision-making that frequently involves patients with diminished autonomy.

Technological advances in medicine introduce more treatment options, and as such they benefit the patients. However, these advances also represent a major challenge for patients and doctors alike; when faced with a myriad of options, the patients may feel confused and resign from making an autonomous decision and instead transfer the responsibility to partners, family members, or doctors. On the other hand, proponents of the concept of "relational autonomy" would argue that no individual makes decisions in isolation and that medical doctors and relatives represent vital and welcome contributors to the clinical decision-making process.

DA should be upheld in all competent patients, who should be provided with all relevant information to a sufficient extent, including information on all available treatment options (where they exist). The patient should express their preference for any or none of the available options voluntarily and without external controlling influences. The patient should also have an option to change their opinion or to transfer the decision-making responsibility to other agents (e.g. family members and doctors). Knowledge of DA criteria enables medical doctors to respect and actively participate in upholding patients' autonomy.

Acknowledgement
This chapter was supported by the Technology Agency of the Czech Republic under the ETA Programme (No. TL05000114).

Bibliography

Arora, Neeraj K., and Colleen A. McHorney. 2000. "Patient Preferences for Medical Decision Making: Who Really Wants to Participate." *Medical Care* 38, 3: 335–341. DOI: https://doi.org/10.1097/00005650-200003000-00010.

Audi, Robert. 1973. "Intending." *Journal of Philosophy* 70, 13: 387–403. DOI: https://doi.org/10.2307/2024677.

Beauchamp, Tom L., and James F. Childress. [1979] 2009. *Principles of Biomedical ethics*. New York: Oxford University Press.

Berg, Jessica W., Paul S. Appelbaum, Charles W. Lidz, and Lisa S. Parker. 2001. *Informed Consent: Legal Theory and Clinical Practice*. Oxford: Oxford University Press.

Borry, Pascal, Paul Schotsmans, and Kris Dierickx. 2006. "Empirical Research in Bioethical Journals. A Quantitative Analysis." *Journal of Medical Ethics* 32: 240–245. DOI: http://dx.doi.org/10.1136/jme.2004.011478.

Brentano, Franz. [1874] 1973. *Psychology from an Empirical Standpoint*. London: Routledge and Kegan Paul.

Charles, Cathy, Amiram Gafni, and Tim Whelan. 1999. "Decision-Making in the Physician-Patient Encounter: Revisiting the Shared Treatment Decision-Making Model." *Social Science & Medicine*. 49, 5: 651–661. DOI: https://doi.org/10.1016/S0277-9536(99)00145-8.

Davies, Rachel, Jonathan Ives, and Michael Dunn. 2015. "A Systematic Review of Empirical Bioethics Methodologies." *BMC Medical Ethics* 16, 15: 1–13. DOI: https://doi.org/10.1186/s12910-015-0010-3.

Dennett, Daniel Clement. 1987. *The Intentional Stance*. Cambridge, MA: The MIT Press.

Eth, Spencer. 1985. "Competency and Consent to Treatment." *JAMA* 253, 6: 778–779.

Fade, Stephanie. 2004. "Using Interpretative Phenomenological Analysis for Public Health Nutrition and Dietetic Research: A Practical Guide." *Proceedings of the Nutrition Society* 63, 4: 647–653. DOI: https://doi.org/10.1079/PNS2004398.

Faden, Rurth R., and Tom L. Beauchamp. 1986. *A History and Theory of Informed Consent*. Oxford: Oxford University Press.

Gert, Bernard, Charles M. Culver, and Danner K. Clouser. 1997. *Bioethics: A Return to Fundamentals*. Oxford: Oxford University Press.

Grisso, Thomas, and Paul S. Appelbaum. 1998. *Assessing Competence to Consent to Treatment: A Guide for Physicians and Other Health Professionals*. New York: Oxford University Press.

Hyman, John. 2013. "Voluntariness and Choice." *Philosophical Quarterly* 63, 253: 683–708. DOI: https://doi.org/10.1111/1467-9213.12074.

Jonsen, Albert R., Mark Siegler, and William J. Winslade. 2015. *Clinical Ethics: A Practical Approach to Ethical Decisions in Clinical Medicine*. New York: McGraw-Hills.

Katz, Jay. 2002. *The Silent World of Doctor and Patient*. Baltimore: Johns Hopkins University Press.

Lesser, Harry. 1983. "Consent, Competency and ECT: A Philosopher's Comment." *Journal of Medical Ethics* 9, 3: 144–145. DOI: http://dx.doi.org/10.1136/jme.9.3.144.

Mental Capacity Act. 2005. Accessed July 15, 2021. http://www.legislation.gov.uk/ukpga/2005/9/pdfs/ukpga_20050009_en.pdf.

Nadelhoffer, Thomas. 2008. *Intentions and Intentional Actions in Ordinary Language and the Law*. Saarbrücken: VDM Verlag Dr. Müller.

Pernick, Martin S. 1982. "The Patient's role in medical decision making: a social history of informed consent in medical therapy." In *Making Health Care Decisions: The Ethical and Legal Implications of Informed Consent in the Patient-Practitioner Relationship, Volume Three: Appendices, Studies on the Foundations of Informed Consent*, ed. by Morris B. Abram, 1–35. Washington, D.C.: The Commission.

Pringle, Jan, John Drummond, Ella McLafferty, and Charles Hendry. 2011. "Interpretative Phenomenological Analysis: A Discussion and Critique." *Nurse Researcher* 18, 3: 20–24.

Ross, Lainie Friedman. 2015. "Theory and Practice of Pediatric Bioethics." *Prespectives in Biology and Medicine* 58, 3: 267–280. DOI: 10.1353/pbm.2016.0008.

Řiháček, Tomáš, Ivo Čermák, Roman Hytych, and al. 2013. *Kvalitativní analýza textů: čtyři přístupy*. Brno: Masarykova univerzita.

Say, Rebecca, Madeleine Murtagh, and Richard Thomson. 2006. "Patients' Preference for Involvement in Medical Decision Making: A Narrative Review." *Patient Education and Counseling* 60, 2: 102–114. DOI: https://doi.org/10.1016/j.pec.2005.02.003.

Searle, John R. 1983. *Intentionality*. Cambridge: Cambridge University Press.

Smith, Jonathan A., and Mike Osborn. 2003. "Interpretative Phenomenological Analysis." In *Qualitative Psychology: A Practical Guide to Research Methods*, ed. by Jonathan A. Smith, 53–80. London: Sage.

Smith, Jonathan A., Paul Flowers, and Michael Larkin. 2009. *Interpretative Phenomenological Analysis. Theory, Method and Research*. London: Sage Publications.

Strull, William M., Bernard Lo, and Charles Gerald. 1984. "Do Patients Want to Participate in Medical Decision Making?" *JAMA* 252, 21: 2990–2994.

Wangmo, Tenzin, and Veerle Provoost. 2017. "The Use of Empirical Research in Bioethics: A Survey of Researchers in Twelve European Countries." *BMC Medical Ethics* 18, 1: 1–12. DOI: https://doi.org/10.1186/s12910-017-0239-0.

Wear, Stephen. 1993. *Informed Consent: Patient Autonomy and Physician Beneficence Within Clinical Medicine*. Boston: Kluwer Academic Publishers.

Chapter 5
Human Cognitive Enhancement and the Problem of Equality

Jana Tomašovičová

Abstract: Visions of human cognitive enhancement are gradually turning into reality thanks to new neurotechnologies, and they have sparked a broad debate on the possible social and ethical implications of this phenomenon. This chapter takes a closer look at the threat of the deepening of social inequalities and the question of how this can be prevented. The first part examines whether the principle of equality of opportunities can be considered a sufficient criterion for judging equality in a given situation. The argument is made that both John Rawls's compensatory measures for the equality of opportunity and their updated version presented by Allen Buchanan et al. have their limitations. For one thing, they do not sufficiently take into account the diversity of human existence (and of human beings) and therefore cannot ensure that no group of people would be excluded from the scope of fairness and equality. In the second part of the chapter, Amartya Sen's and Martha Nussbaum's capability approach is analysed with regard to whether it is able to eliminate new forms of discrimination and exclusion that may arise as a result of cognitive enhancement. This discussion includes the possibility of coping with demands for the recognition of the equality of new and enhanced forms of life. It is argued that the capability approach is a more complex and differentiated conceptual framework for thinking about equality in the context of human cognitive enhancement than what is provided by Rawls's theory of justice for at least two reasons.
Keywords: Cognitive enhancement, resource distribution, equality of opportunity, capability approach, diversity of human existence, recognition of equality of enhanced life forms.

Introduction

Human enhancement is becoming increasingly real thanks to the rapid development of new technologies. Current convergent technologies seek to uncover hitherto unknown dimensions of the human body and identify its basic structures by effectively combining research methods and knowledge from different scientific disciplines, thus extending and deepening the existing knowledge of human beings in many ways; however, they also create room for the phenomenon of human enhancement in combination with the ancient human desire to improve limited natural abilities – i.e. the purposeful expansion and intensification of physical, cognitive, emotional, and even character traits.[60] In contrast to humanistic forms of enhancement – such as education, training, and upbringing – human enhance-

[60] For the specific use of the concept of "enhancement" in the field of bioethics, see Schöne-Seifert and Stroop (2015, 249). A more detailed analysis of neurotechnological methods that can be used for the purposes of enhancement is given by Clausen (2008, 39–58).

ment involves the use of the latest technologies. The technological modification of the human body and the active intervention in its biological processes signifies a gradual blurring of boundaries between nature and culture.

As a result of these research trends, a number of interpretative schemes and standards of assessment are gradually losing their validity and are ceasing to provide sufficient guidance in dealing with newly emerging issues. Dieter Birnbacher points out that the standards of assessment are mainly conceptual tools used to organize and explain regularly occurring phenomena; therefore, they cannot be expected to work reliably in every newly emerging situation. They may turn out to be unusable or insufficiently "sharp"; in new contexts, it will probably be necessary to test their validity and search for more adequate solutions (Birnbacher 2006, 281–282).

The object of reasoning in this chapter is the examination of selected concepts and principles that are most commonly used in assessing the expected social consequences of cognitive enhancement. Potential medical, ethical, and social issues are examined in the context of human cognitive enhancement, with issues of equality and distributive justice dominating the discussion of the social consequences.[61] In the first part, this chapter considers how the exacerbation of social inequalities in the context of cognitive enhancement can be avoided and whether the principle of equality of opportunity can be considered a sufficient criterion for judging equality in a given situation. In the context of cognitive enhancement, concerns about possible discrimination against non-enhanced people have to be taken into consideration. The issue of equality takes on a broader dimension in this context, and the second part of the chapter therefore addresses the question of whether existing concepts of equality are capable of eliminating new forms of discrimination and exclusion that might arise as a result of cognitive enhancement, and whether they include the possibility of coping with demands for the recognition of equality for new and enhanced lifeforms.

Equal access and the distribution of resources

The social problems associated with cognitive enhancement relate primarily to issues of equality and distributive justice. The starting point for this is based on the assumption that cognitive enhancement will enable its users to grow in competence, thereby increasing their advantages in competing for job opportunities.

[61] For interdisciplinary perspectives reflecting the diverse aspects and potential implications of neuroenhancement, see Schütz, Hildt, and Hampel, eds. (2016); Viertbauer and Kögerler, eds. (2019); and Sýkora (2019, 511–529). For analyses of individual autonomy, social pressure, and fair access due to neuroenhancement, see Tomašovičová (2021, 181–194).

Since enhancement (unlike therapeutic procedures) is not supposed to be covered by public health insurance, which is primarily intended to cover the costs of curing and treating diseases, it can be thus assumed that it would not be equally available to everybody. The already disadvantaged lower social class would not be able to afford it. Unequal access to enhancement may thus lead to an inequality of opportunity, and there are growing concerns about increasing social inequality and the widening of socioeconomic disparities.

In this context, advocates of cognitive enhancement argue that society already accepts private education and supplementary courses that only the children of well-off parents can afford. This qualitatively superior type of education significantly expands children's cognitive abilities and improves their initial conditions for employment (Caplan 2009, 165–168). Inequality in the form of unequal access to the acquisition of cognitive abilities is already present in society; according to Arthur Caplan, there is no fundamental difference between the "enhancement" acquired through an exclusively private education and technical enhancement (2009, 167). If we accept the former despite there being unequal access to it, why should we disqualify the latter for the same reason? The possible disadvantage of the underprivileged is therefore not a reason to prohibit or restrict enhancement but rather a stimulus to correct existing developments and their effects on the disadvantaged.

The mechanism of correction to eliminate initial social inequalities and ensure the equality of opportunity was discussed by John Rawls in his theory of justice. This mechanism of correction has more recently been adopted by a number of advocates of enhancement. Rawls dealt primarily with the fair equality of opportunity: "[F]air equality of opportunity is said to require not merely that public offices and social positions be open in the formal sense, but that all should have a fair chance to attain them" (2001, 43). This means that it is not sufficient to simply formally declare equal rights to education and social positions; it is necessary to also ensure fair accessibility to them. Factors affecting equality of opportunity, and which enter the game as its preconditions, must also be taken into consideration. According to Rawls, these factors are mainly social and natural in nature. This concerns the social origins and status of the families into which people are born and which they grow up in alongside the biological preconditions that are manifested in the diversity of their talents and physical qualities (Rawls 2001, 55). Rawls considers these factors to be morally arbitrary, because no individual has personally contributed to them.[62] Since even in a well-ordered society they tend to cause problematic inequalities, they cannot be ignored; a system of regulations must therefore be established to help eliminate this natural "lottery" (Rawls 2001, 56). According to Rawls, it would be unfair if equally or similarly talented individuals were less likely to succeed and further develop their talents simply because they came from inferior social backgrounds. He proposes a system of compensation that would ensure equality of starting conditions in education and employment: "[A]ssuming that there is a distribution of natural assets, those who are at the same level of talent and ability, and have the same willingness to use them, should have the same prospects of success regardless of their initial place in the social system" (Rawls 1999, 63). In order for a society to avoid increasing social inequalities, Rawls proposed the introduction of a system of compensation in the form of equalizing initial opportunities.

If cognitive enhancement – as a technological or pharmacological modification of natural biological dispositions – enters into this situation of rule setting for the fair functioning of society, it is necessary to take this factor into account as something with a real impact on the equality of opportunity; however, the possibility of state support that would mitigate unequal access to enhancement, as in the case of equalizing educational opportunities, interferes with one of the pillars of liberal theory – namely, the state's neutrality in relation to different individual and partial conceptions and preferences for a good life. Supporting only those goods which are generally necessary for people to develop adequately as members of society and to realize their life ambitions would be compatible with neutrality. Rawls defines these as "primary goods", and he argues they should be distributed to one and all (Rawls 2001, 58–59).

In response to the newly emerging situation associated with cognitive enhancement, and analogous to Rawls's idea of "social" primary goods, Allen Buchanan, Dan Brock, Norman Daniels, and Daniel Wikler developed the argument that a person's cognitive abilities can be considered a "natural" primary good because cognitive abilities are necessary for the realization of an individual's ideas and life plan and are thus important for the successful implementation of practically every life project (Buchanan et al. 2009, 278–281).[63] The loss or lack of such abilities threatens almost all life plans. According to Buchanan et al., cognitive abilities are general purpose means – i.e. means which are necessary for every

[62] "Do people really think that they (morally) deserve to be born more gifted than others?" (Rawls 2001, 74). According to Rawls, the distribution of innate ability is undeserved because "moral desert always involves some conscientious effort of will, or something intentionally or willingly done" (74, note 42).
[63] Also see Buchanan et al. (2000) and Buchanan (1995).

purpose. Upon this basis, it can be concluded that, if necessary, an appropriately set social programme could support the enhancement of cognitive abilities in people who are socially disadvantaged. This would regulate and equalize their starting opportunities. Even if one accepts Rawls's assertion that natural biological dispositions are not morally meritorious, as they are not the results of individual endeavour, supporting cognitive enhancement of the less talented can correct the impact of the natural lottery. Once cognitive enhancement is launched, the theory of the widening of socioeconomic disparities need not be fulfilled. Rather, it could be prevented by supporting the disadvantaged while not restricting the privileged (Galert et al. 2009, 8).

A number of criticisms must, however, be made concerning the reasoning outlined above. Firstly, there is no getting around the fact that the principle of equal opportunities itself has certain limits. For one thing, it does not sufficiently take into account those people who are unable to grasp and take advantage of the equality of opportunity due to various limitations and disabilities that are not of their own making. Disabled, sick, and elderly people are more likely to have special needs and demands resulting from various health and biological factors. They also require a guarantee of "special opportunities" in order to lead a valuable and dignified life. Despite this, they are not explicitly dealt with in Rawls's theory of justice and they are not part of compensatory measures. The proposed principle of equality of opportunity thus does not function as an adequate criterion of equality.

Secondly, the principle of equality of opportunity is unlikely to be a sufficient criterion of equality for those who refuse improvements for various reasons. The risks of their possible discrimination and the potential sources of associated tension in society should not be underestimated and left unnoticed. They pose a challenge in the search for effective tools to regulate possible inequalities caused by cognitive enhancement.

Thirdly, there is a failure to fully ensure the fair equality of opportunity: even under current circumstances, where significant income disparities are tolerated. This increases the chances for certain individuals to socially benefit from their position. Increased caution in introducing technical advancements due to their potential risks of widening inequality is therefore justified and fully legitimate given the overall functioning of society.[64] This is clearly not a fundamental

[64] In a broader context, Andrej Démuth and Slávka Démuthová (2020, 50–62) also reflect on the need to strengthen and restore public trust in justice.

reason to restrict research in the field of cognitive enhancement; however, it is a good reason to carefully examine its possible implications.

The above suggests that a solution to unequal access to enhancement, relying in particular on Rawls's argument and its updated version (Buchanan et al. 2000), could be found in the introduction of a system of compensatory measures to even out unequal initial social conditions; however, such a proposal does not answer the question of the situation of people who are medically disadvantaged and for whom guaranteeing equality of opportunity does not constitute a sufficient solution due to their increased and legitimate demands for dignified functioning. It also leaves open the question of whether people from disadvantaged social backgrounds – for whom the compensation would be intended – would refuse enhancement for various reasons rather than opting for it. How can there be an assurance that no groups of people would be excluded from the scope of fairness and equality? How can the potential for discrimination be prevented and the preconditions for a two-class society be eliminated? Starting from this broader context, it appears that equality is a much more complex issue and should not be reduced to a simple matter of unequal access to enhancement. In the next section, this chapter shall examine whether a method based on the assumption of the diversity of human existence – and which does not aspire to reduce or overlook this diversity in any way – is a more appropriate framework for considering equality in the context of human enhancement.

The equality of what?

Given such reservations, all forms of human existence must be taken into account in setting the rules for a justly functioning society. Reasoning cannot be narrowed – as Rawls did – to subjects who are fully autonomous and rational, and who "under the veil of ignorance" (Rawls 1999, 118) can clearly articulate and defend their own interests. The disabled, the socially excluded, the sick and elderly, and, soon enough, even the enhanced and unenhanced are all legitimate parts of society. In other words, various forms of human existence must be taken into account when considering equality and justice. If one was to start from this assumption of human diversity, then this results in the insufficiency of the criteria that are aimed at ensuring an equality of opportunity or an equality of primary goods; the capacities to convert these acquired goods into a valuable way of being substantially vary for different forms of human existence (Sen 1980, 219). The Indian economist and philosopher Amartya Sen was one of the first to draw attention to this fact when he proposed assessing equality in terms of basic human capabilities –

i.e. the real possibilities and freedoms of individuals to achieve valuable social functioning (Sen 1992, 40).

Arguing against Rawls's theory of justice, Sen asserts that the index of primary goods is not a sufficient measure of equality. Even though primary goods are conceived quite broadly and inclusively – as they include basic rights and freedoms, opportunities, income, wealth, and the social foundations of self-esteem – Sen states that attention should not be placed on the goods themselves. Nonetheless, a more essential aspect that is absent from Rawls's approach is the focus on the relationship between goods and people. This means observing whether the goods in question actually enable people to lead worthwhile and dignified lives (Sen 1980, 216). This aspect is important for the reason that people are very different (219). Given their health, age, intelligence, social conditions, and other conversion factors, their ability to use abstract resources in order to achieve realistic opportunities to live and function with dignity and value varies substantially. The same distribution of resources would necessarily be inadequate for people with disabilities who have legitimate increased demands and needs due to their illnesses (Bickenbach 2014, 12). Taking into account the various forms of human existence, Sen asserts that resources and primary goods on their own cannot be a sufficient indicator of equality and justice. Resources should not be the objective of society's efforts but rather a means to valuable goals.

The main reason for Sen's critique of Rawls's theory of justice was the lack of the consideration of human life in its plurality of forms. He obligingly notes that if people were very similar, then Rawls's fair distribution of primary goods – and guaranteeing the equality of opportunity – could presumably function as an adequate measure of equality. Interpersonal differences, however, are now so significant that overlooking them leads to a partially blind morality (Sen 1980, 216). In the context of possible enhancements in human cognitive abilities, it is reasonable to assume that these interpersonal differences will continue to grow significantly. The question of determining the relevant criterion of equality, especially in light of the possible widening of inequalities and differences between people in the near future, is therefore a fully legitimate one for a justly functioning society.

The question then is as follows: If it turns out that neither the equality of resources or primary goods (egalitarianism) nor the equality of opportunity are sufficient criteria for equality when taking human diversity into account, what other (more appropriate) criterion can be considered? Sen observes that the prerequisite for valuable social functioning – the precondition for a good life for any form of human existence – is its capabilities: i.e. the actual possibilities of leading a dignified life. These capabilities represent the real possibilities of a human being and the freedom of that human being to do and be what they have a reason to value (Sen 1992, 40; 1980, 218). These possibilities are created by a combination of internal and individual preconditions (e.g. health, age, and talent) as well as external (social, economic, political, and environmental) ones alongside other factors. Naturally, the spectrum of capabilities is vast; not all of them are equally important, which is why Sen proposes assessing equality by taking into account individuals' basic capabilities. These are capabilities that can be considered essential for a dignified life and that enable a person to avoid poverty, deprivation, and conditions unworthy of a dignified life (Sen 1992, 45; 1980, 218).[65] The specification and particular definition of these basic capabilities should be decided by each society or culture in the form of an open public discourse. This would indicate what a given society considers to be the necessary conditions for achieving a worthy and dignified life. Sen does not create a universal "theory" of justice as such but rather identifies a conceptual framework that allows for the assessment of the extent of human inequalities, poverty, and deprivation in real time and space, and which proposes specific social measures to eradicate them. This framework is defined by two poles: capabilities and function (meaning the real fulfilment of the capabilities). When assessing equality, the focus is on core capabilities. Out of a given set of capabilities, what an individual undertakes and fulfils is the result of their own free choice (Sen 1992, 49).[66]

Unlike Sen, the philosopher Martha Nussbaum has attempted to directly identify a list of ten central human capabilities that she considers to be constitutive of a dignified human life and which she presents as a sufficient means of measuring social justice. Nussbaum refers to the dignity of life in terms of Aristotle's concept of the good life ("human flourishing"). This means that human beings are guaranteed certain basic conditions for survival and dignified living. These conditions, termed by Nussbaum as "central capabilities", are so essential that without them human life would be seriously impoverished (Nussbaum 2011, 31).

[65] Giorgio Agamben also draws attention to the need for increased caution in assessing human life. This cautionary note is particularly important so that the mistakes of the past are not repeated and so that society does not slip back into distinguishing between those lives that are worthy of living and those that are not (Agamben 2002).

[66] The United Nations Development Programme has used a capability-based approach in the design of its annual Human Development Reports. This approach provided a broader framework for assessment, emphasizing the expansion of human opportunities and freedoms in achieving worthwhile goals, thus providing a balance to narrowly defined economic indicators (Robeyns 2006, 351). For a more detailed discussion on the multiple dimensions of human development, see Alkire (2002, 181–205).

These central capacities must be seen as mutually irreplaceable. They are all equally important, and one cannot be a substitute for another. According to Nussbaum, these are: "life; bodily health; bodily integrity; senses, imagination, and thought; emotions; practical reason; affiliation (interpersonal association and the social bases of self-respect); other species; play; [and] control over one's environment (political and material)" (Nussbaum 2006, 76–78; 2011, 33–34). They are deliberately formulated in an abstract manner to make it clear that a basic normative framework for a decent and just society is necessary; at the same time, this framework must remain flexible and open for possible additions, further specifications, and revisions based upon cultural particularities and social consensus.[67] The essential consideration is that the list provides a philosophical basis for a just society which should at least guarantee its citizens a threshold level of each capability through constitutional means (Nussbaum 2006, 71). Given that these central capabilities are necessary conditions for a dignified life, Nussbaum asserts that they can therefore be interpreted as basic claims made by human beings in relation to the state; indeed, they form a partial and minimal account of social justice (Nussbaum 2006, 71).

The above analysis clearly shows why the state should guarantee such necessities to its citizens. Nussbaum's argument is the principle of the dignity of every member of society. The principle of dignity – expanded by the Aristotelian dimension of the practical capability to lead a worthwhile life – requires that every person be guaranteed a set of basic entitlements necessary for social functioning. At the same time, the capability approach implies that the issue of equality cannot be linked solely to the equality of resources and primary goods, or indeed to the equality of opportunity. Neither the equality of resources nor the equality of opportunity can guarantee a valuable way of being and social functioning for everyone. A more differentiated approach in assessing equality is needed to ensure that none of the aspects of human diversity are omitted.

In the context of human enhancement and the ongoing debate on the social implications of this phenomenon, the question of how a capability approach can contribute to this debate presents itself. This is even considering the fact that this deals with which fundamental pillars of society should be preserved and which should be rethought and rebuilt. In the context of human enhancement and its possible consequences for individuals and society, the capability approach is a more differentiated conceptual framework than Rawls's theory of justice and

[67] Johann Roduit, Jan-Christoph Heilinger, and Holger Baumann examine the possibility of using central capabilities as a basic referential framework for guiding human enhancement (2015, 622–630). Such an interpretation of central capabilities is questioned by Ivars Neiders (2019, 85–102).

allows previous considerations of equality to be extended for at least two reasons. Firstly, it takes into account human life in its various forms, and it creates the right conditions for eliminating diverse forms of discrimination and social exclusion; this refers to current forms as well as those forms that may arise in the near future, particularly in relation to unenhanced people. If a capability approach emphasizes the provision of basic capabilities in terms of fundamental legal rights for every member of society, it can reasonably be assumed that this will help to create and cultivate a social environment that is suitable for any form of human existence and which removes elements of the potential discrimination or stigmatization of the most vulnerable groups.

Secondly, in the context of human enhancement, it is expected that there will be increasing demands in society for the recognition of new and enhanced transhuman and posthuman life forms. Given the perspective of Rawls's theory of justice – where primary involvement was by autonomous and rational subjects in the compilation of the conditions of society's functioning and the formulation of the criteria of coexistence based upon their own preferences and interests – it may be somewhat problematic for these subjects to accept and recognize the equality of completely different and enhanced beings in a given society. This is problematic in the same way when incorporating the disabled and sick into Rawls's principles of justice. Nonetheless, if one looks at this situation from a capability approach, which respects human diversity at the outset, then it can be assumed that it will also provide sufficient room for the expansion and recognition of new kinds of equality. The coexistence of enhanced and unenhanced forms of life is very likely to be one of the key issues for society in the near future. Sen and Nussbaum's emphasis on diversity very much corresponds with the vision of the transhumanist Nick Bostrom, who argues that different types of existence with different enhancements will coexist side by side in the near future (Bostrom 2018). According to Bostrom, the existence of different forms within a society does not automatically imply the breakdown of society or slavery but rather the need for a more intensive search for effective social solutions with respect to the newly emerging conditioning factors (2018, 97). Just as contemporary society is struggling to find and apply effective protective and regulatory mechanisms to redress inequalities, society in the future will face a similar task. Meanwhile, the capability approach has sufficient potential to function as a conceptual framework, even in the case of a new configuration of social relations in which social measures will be set up to prevent deprivation, respect diversity, and provide minimum basic capabilities to all diverse forms of existence.

Conclusion

Given the central role that the human brain plays in life, it is understandable that current debates have intensively analysed the possibilities and the risks of human cognitive enhancement from multiple perspectives. This chapter focused on exploring the social implications of cognitive enhancement, considering in particular the possibilities for avoiding the potential deepening of social inequality. It relied on two concepts – Rawls's theory of justice and Sen's capability approach – and explored the extent to which they can cope with emerging issues of equality. Rawls's proposal for compensatory measures to redress initial social inequalities and ensure the equality of opportunity was expanded upon by Buchanan et al. in an innovative way. This, however, precisely reflects the enhancement of cognitive abilities, and even the extended and updated proposal shows some limitations. For instance, it does not take into account the diversity of human existence and thus overlooks the fact that the proposed system of compensation for equalizing opportunities is insufficient for the disabled, the sick and elderly, and (in the near future) probably also the unenhanced.

A more appropriate framework for contemplating equality issues in the context of cognitive enhancement appears to be the capability approach, and this is primarily for two reasons. Its initial consideration of the diversity of human existence creates appropriate conditions for eliminating the various forms of discrimination and stigmatization of vulnerable groups in society, including people who, for various reasons (even reasonable ones), will refuse enhancement. At the same time, it creates the right conditions for the recognition of new forms of equality, which is very likely to be one of the key social issues in the near future. In this way, the capability approach is a more comprehensive and differentiated conceptual framework for thinking about equality in the context of the anticipated expansion of human cognitive enhancement. Nonetheless, contemporary philosophy needs to keep a close eye on these developments and respond to them as needed by examining the relevance of explanatory concepts that have been valid thus far and identifying new ways of assessing them.

Acknowledgement

This chapter was written at the Department of Philosophy and Applied Philosophy, Faculty of Arts, University of Ss. Cyril and Methodius in Trnava, as a part of VEGA project 1/0563/18.

Bibliography

Agamben, Giorgio. 2002. *Homo sacer. Die souveräne Macht und das nackte Leben*. Trans. by Hubert Thüring. Frankfurt am Main: Suhrkamp Verlag.

Alkire, Sabina. 2002. "Dimensions of Human Development." *World Development* 30, 2: 181–205.

Bickenbach, Jerome. 2014. "Reconciling the Capability Approach and the ICF." *Alter, European Journal of Disability Research* 8, 1: 10–23. DOI: https://doi.org/10.1016/j.alter.2013.08.003.

Birnbacher, Dieter. 2006. *Bioethik zwischen Natur und Interesse*. Frankfurt am Main: Suhrkamp Verlag.

Bostrom, Nick. 2018. *Die Zukunft der Menschheit. Aufsätze*. Trans. by Jan-Erik Strasser. Berlin: Suhrkamp Verlag.

Buchanan, Allen E. 1995. "Equal Opportunity and Genetic Intervention." *Social Philosophy and Policy* 12, 2: 105–135. DOI: 10.1017/S0265052500004696.

Buchanan, Allen E., Dan W. Brock, Norman Daniels, and Daniel Wikler. 2000. *From Chance to Choice: Genetics and Justice*. Cambridge: Cambridge University Press.

Buchanan, Allen E., Dan W. Brock, Norman Daniels, and Daniel Wikler. 2009. "Warum nicht das Beste?" In *Enhancement. Die ethische Debatte*, ed. by Bettina Schöne-Seifert and Davinia Talbot, 267–294. Paderborn: Mentis Verlag.

Caplan, Arthur L. 2009. "Ist besser das Beste? Ein renommierter Ethiker plädiert für Enhancement des Gehirns." In *Enhancement. Die ethische Debatte*, ed. by Bettina Schöne-Seifert and Davinia Talbot, 165–168. Paderborn: Mentis Verlag.

Clausen, Jens. 2008. "Gehirn-Computer-Schnittstellen: Anthropologisch-ethische Aspekte moderner Neurotechnologien." In *Die "Natur des Menschen" in Neurowissenschaft und Neuroethik*, ed. by Jens Clausen, Oliver Müller, and Giovanni Maio, 39–58. Würzburg: Königshausen & Neumann Verlag.

Démuth, Andrej, and Slávka Démuthová. 2020. "Confidence in Justice as a Moral Emotion and Five Mechanisms that Support its Renewal or Enhancement." *Právny obzor* 103, special issue: 50–62.

Galert, Thorsten et al. 2009. "Das optimierte Gehirn." *Gehirn & Geist* 11. Accessed June 7, 2021. https://www.spektrum.de/sixcms/media.php/976/Gehirn_und_Geist_Memorandum.pdf.

Neiders, Ivars. 2019. "Can We Use the Capabilities Approach to Evaluate Human Enhancement?" In *Promises and Perils of Emerging Technologies for Human Condition: Voices from Four Postcommunist Central and East European Countries*, ed. by Peter Sýkora, 85–102. Berlin: Peter Lang.

Nussbaum, Martha C. 2006. *Frontiers of Justice: Disability, Nationality, Species Membership*. Cambridge: The Belknap Press of Harvard University Press.

Nussbaum, Martha C. 2011. *Creating Capabilities: The Human Development Approach*. Cambridge: The Belknap Press of Harvard University Press.

Rawls, John. [1971] 1999. *A Theory of Justice*. Cambridge: The Belknap Press of Harvard University Press.

Rawls, John. 2001. *Justice as Fairness: A Restatement*. Cambridge: The Belknap Press of Harvard University Press.

Robeyns, Ingrid. 2006. "The Capability Approach in Practice." *The Journal of Political Philosophy* 14, 3: 351–376.

Roduit, Johann A. R., Jan-Christoph Heilinger, and Holger Baumann. 2015. "Ideas of Perfection and the Ethics of Human Enhancement." *Bioethics* 29, 9: 622–630. DOI: https://doi.org/10.1111/bioe.12192.

Schöne-Seifert, Bettina, and Barbara Stroop. 2015. "Enhancement." In *Handbuch Bioethik*, ed. by Dieter Sturma and Bert Heinrichs, 249–254. Stuttgart, Weimar: J. B. Metzler Verlag.

Schütz, Ronja, Elisabeth Hildt, and Jürgen Hampel (eds.). 2016. *Neuroenhancement. Interdisziplinäre Perspektiven auf eine Kontroverse*. Bielefeld: Transcript Verlag.

Sen, Amartya. 1980. "Equality of What?" In *The Tanner Lectures on Human Values*, ed. by Sterling M. McMurrin, 195–220. Salt Lake City: University of Utah Press; Cambridge: Cambridge University Press.

Sen, Amartya. 1992. *Inequality Reexamined*. Cambridge: Harvard University Press.

Sýkora, Peter. 2019. "Towards the Posthuman Through the Editing of Genes for Cognitive Capabilities." *Filozofia* 74, 7: 511–529. DOI: https://doi.org/10.31577/filozofia.2019.74.7.1.

Tomašovičová, Jana. 2021. "Social and Ethical Consequences of Neuroenhancement." *Filozofia* 76, 3: 181–194. DOI: https://doi.org/10.31577/filozofia.2021.76.3.2.

Viertbauer, Klaus, and Reinhart Kögerler (eds.). 2019. *Neuroenhancement. Die philosophische Debatte*. Berlin: Suhrkamp Verlag.

Chapter 6

Prometheus the Biohacker? Mythical Grammar in the Discourse of Bioscience After the CRISPR Revolution

Mariusz Pisarski

Abstract: The ethics of gene editing is a highly contested space where different disciplines and voices have different things to say about what should be publicly acceptable with regard to gene therapy, its accessibility, and the limits that should be imposed on its use. Such a contest is taking place in a discursive field where boundaries between fact and fiction are more blurred than ever. Additionally, developments in biotechnology are so rapid that in order to describe them, both commentators and scientists refer to science fiction. The goal of this chapter is to demonstrate that an additional repertoire of interdisciplinary language can be found in science fiction as well as in Greek mythology. I will reflect on science fiction motifs and current discussions on DIY bioengineering and gene therapies as a form of the contemporary enactment of the myth of Prometheus. To emphasize the blurring of discursive boundaries, visions of the near future from cyberpunk and biopunk narratives will also be contrasted and compared with the contemporary discourse on the psychological and socio-economic impact of biotechnology. The fictional sources of reflection include the computer game *Cyberpunk 2077* by CD Project Red (2020) and the biopunk fiction novel *The Windup Girl* by Paolo Bacigalupi (2009). The non-fictional material is supplied by the documentary series *Unnatural Selection: Is Biohacking Ethical?* (2019). The methodology of the chapter blurs discursive boundaries by drawing from semiology and narratology on the one side and the general discourse of bioscience (areas of bioethics, biopolitical discourse in arts, and DIY science) on the other. The comparative study of mythical motifs in fictional and non-fictional visions of the future of human gene editing aims to deliver a cultural context to the issue of the growing gap between science and anti-science, knowledge and conspiracy theories, and scientific progress and corporate interests.
Keywords: Myth, semiology, posthumanism, transhumanism, cyberpunk, biohacking, biopolitics.

Introduction

According to Mircea Eliade, one of the fundamental functions of myths is to establish models for behaviour (Eliade 1963, 2). Such a behaviour-generating role is especially useful when facing the unknown. In these situations, myths allow members of society to interpret and fit unfamiliar situations into old and familiar frames, construct a “language of argument”, and organize reality and experience into recognizable patterns (Breen and Corcoran 1982, 17). Such myths can be quite useful today for many areas and disciplines, especially in those areas where the pace of change is so fast and the consequences are so hard to predict. Bioscience (and bioengineering) might be best suited for the inclusion of myths, especially when the social, economic, and ethical consequences of their discoveries are the focus of interdisciplinary reflection. Debates on bioengineering have become inevitably heated, especially after the "CRISPR revolution" which brought cheap and accessible tools for human genome editing, and myths can greatly contribute to the discussion. J. B. S. Haldane, a visionary biologist and the author of *Biological Possibilities for the Human Species in the Next Ten Thousand Years* (1963) encouraged turning towards myths in the context of scientific progress in understanding, "deconstructing", and taking control of evolution. The chemical or physical inventor, Haldane argued, is always a Prometheus:

> There is no great invention, from fire to flying, which has not been hailed as an insult to some god. But if every physical and chemical invention is a blasphemy, every biological invention is a perversion. (Haldane 1995, 36)

Haldane's remarks point to a resistance towards science. However, other forms of resistance – directed not towards the invention itself, but towards those who are in control and possession of it – are of equal importance. This sort of resistance is particularly visible when discussing gene editing and access to gene therapies. Dystopian fiction of the near future, especially in the cyberpunk and biopunk genres, envisions a future where the fruits of biotechnology, such as human enhancement and extended longevity, are not evenly distributed and are controlled by corporations. This, in turn, functions as a narrative trigger and sets protagonists on a path of resistance against such post-governmental forms of biopower.

According to Michel Foucault, resistance is the key word and prime impulse of modern power (Foucault 2019, 167). Contemporary scholars have extended the importance of resistance as an element that precedes power to the notion of biopower (Lazaratto 2002, 122) and into the world of emerging biotechnologies and issues of the production, distribution, and consumption of resources (Thacker and Gerring 2008, 310–311). The phenomenon of resistance becomes central to discussions about DIY science and the positioning of the amateur scientist, artist, and activist in this discursive field (Pentecost 2008, 113). One can identify two main targets of this resistance. The first of these is formed by centres of power within the field in question; in the case of biotechnology, these are people and institutions with a decisive role in the flow of knowledge and resources. To borrow Pierre Bourdieu's terminology, these are the main agents within the field (Bourdieu 2005, 193). The second source of power structures can be found in the very notion of society (or the understanding of human nature) as something that is controlled by our DNA (Lewontin 1996, 61). In other words, those who represent resistance – DIY scientists, citizen scientists, and biohackers – are not only against Big Science but are also against a political philosophy that makes human nature unchangeable and coded in our genes. They want to change that by putting the tools of bioscience "in the hands of everyone who wants them" (Patterson 2010) and making the results of genetic engineering available to everyone at a low cost. Often, these efforts are made in the spirit of the "creative evolution" and a person's right to their own body.

The aim of this chapter is to look at the notion of resistance from the perspective of the Promethean myth and to present the myth of the Titan who steals fire and *techne* from the gods to enrich humanity. Myths can function as a cultural reservoir of potent ideas, images, and vocabulary that are able to influence behaviour and accommodate discussions on the social, ethical, and economic consequences of genetic engineering. How do concepts of individual autonomy and freedom – and the notion of progress – change in the context of modern bio-scientific research? To what extent is the scientific community and public opinion ready for curbing some freedoms in order to control the possible (and not always predictable) consequences of gene editing? Although no obvious answers exist to these questions, the scientific community is in need of developing some common discursive denominators that would help participants engaged in the discussion effectively communicate with each other within the emerging "biodiscourse", where concepts and methods of several disciplines – such as science, art, and philosophy – merge. Comprehending current developments in biotechnology and the relationships between the field's main actors as occurrences of myths – of Prometheus, Frankenstein, and the Mad Scientist – can be a way to bring order and structure to heated discussions of an ethical, political, and religious nature.

I will base my reflection on examples drawn from the computer game *Cyberpunk 2077* (CD Project RED), the biopunk science fiction novel *The Windup Girl* by Paolo Bacigalupi (2009), and the Netflix documentary series *Unnatural Selection* (2019). The grouping of fictional and non-fictional material for this study is by no means accidental. Within any discourse, the power of words and ideas, even if they relate to fictional entities, can be of equal performative potential as the power of real-life events. The former often influence the latter. This pattern is especially clear in the discourse of bioscience, where literature – and speculative fiction and science fiction in particular – can emerge as a generative site where art, literature, culture, and politics converge (Cardozo and Subramaniam 2008, 269).

In the reflection, the importance of *logos*[68] for the future directions of the developing biodiscourse is reinforced by the emphasis on *mythos*. According to Roland Barthes, who in his *Mythologies* (1957) pioneered a semiology of myth in everyday life, mythic structures permeate every act of cultural communication: from art to advertisements and from elaborate forms to a single photograph on the cover of a weekly magazine (Barthes 1991, 142). Under the semiotic mechanism described by Barthes, the linguistic sign is turned into a mythical signifier. As a result, the visible meaning of the first order (i.e. what is perceived in the message) is turned into the meaning of the second order, which uncovers a myth. In a similar manner, this study focuses on events and characters (both fictional and non-fictional) whose words and actions trigger mythical stories of progress, emancipation, and rebellion against the established structures of biopower. Although the characters themselves may never explicitly refer to mythical motifs, their actions instantaneously trigger such second-order meanings, which are ready to be analysed and compared. These uncovered mythical structures are worth studying because they are written in a universal language that is spoken across disciplines and across cultures – something that might prove essential for effective communication within biodiscourse – and also because (to invoke Sigmund Freud and his methods of psychoanalysis) second-order meaning is the "ultimate meaning" of human behaviour (Freud 2010, 365).

The Promethean myth of enhancement and progress

According to the Platonic retelling of the myth of genesis in Protagoras, the gods left the human species unequipped and weak in comparison to other animals. After the creation, two Titans – Prometheus and his brother Epimetheus – were in charge of equipping the creatures with "powers" (Plato in Arieti and Barrus 2010, 55). Epimetheus was so absorbed in the tasks of attributing the gifts of speed, agility, good eyesight, hearing, and suchlike to nonreasoning creatures that in the end there was nothing left for the naked and defenceless humans. At this stage, Prometheus decided to compensate for Epimetheus's oversight and aid humans in their struggle for survival. To do so, Prometheus stole the powers of technical skills, speech, and fire from the gods and gave them to the first humans (Mayor 2018, 61). For this, Prometheus was punished by Zeus and chained to a rock in the Caucasus Mountains. Recounted by Aristophanes in his tragedy *Prometheus*

[68] Logos is understood here as a way of structuring and rationalizing an argument by linguistic expression. According to Barthes, *mythos* can be seen as a logos of logos, a type of metalanguage that borrows its second order of meaning from the pre-existing acts of communication. As such, it is also regarded by Barthes as "depoliticized speech" (Barthes 1991, 142) and a "stolen language" (Barthes 1991, 131).

Bound and retold ever since, with famous examples from the Romantic period such as Lord Byron's *Prometheus* (1816) and Percy Bysshe Shelley's *Prometheus Unbound* (1820), the myth of Prometheus is deeply engrained in any discourse on humans' relationships to technology, human enhancement, and scientific progress. Mary Wollstonecraft Shelley's *Frankenstein; or, the Modern Prometheus* is directly connected to Percy Shelley's Romantic rendition of the myth with an emphasis on sacrifice for the betterment of humanity. Created almost concurrently with Percy Shelley's famous poem in 1818, *Frankenstein* launched the Promethean myth into a new era and new discourse which has bound these motifs closely to modern science, technology, and the ethical obligations of scientists. As modern gods, scientists are able to create a "new species" of man (Shelley 1999, 43), a power that since then has been one of the central motifs of both science fiction and public discourse surrounding the social, economic, and ethical dimensions of scientific progress.

Establishing the framework for future debates about the relationship between man and technology, Mary Shelley's seminal work initiated ever-lasting debates over which Promethean aspects of science and scientists were discussed in a range of contexts from the modern to the postmodern and the posthuman (Rogers 2018, 206–227). In these accounts, scientists are elevated to the level of the Titans. However, they are presented on a moral spectrum marked by figures of a benevolent genius and a rebel to a demented demiurge (Gomel 2011, 343) among other popular depictions.

Although recurrences of Promethean motifs in literature and popular culture have been extensively researched, with one of the latest contributions made by the Bloomsbury monograph *Frankenstein and Its Classics: the Modern Prometheus from Antiquity to Science Fiction* (Weiner, Stevens, and Rogers 2018), the reflection on their presence as a second-order meaning in fictional and non-fictional accounts of the application of biotechnology and genetic enhancement on the human body, especially since the emergence of CRISPR/Cas9 technology, has not been discussed very much. This chapter aims to fill this gap by examining structures of biopower, resistance against it, and biopolitical configurations present in science fiction and games and the contemporary biohacker scene of DIY genetic engineering. An important shift in the approach to the Promethean myth by *Frankenstein* and later expressions of the theme also needs to be noted. Scholars agree that Shelley's subtitle of "the Modern Prometheus" should be seen as ironic and part of the novel's polemic with both Romantic Prometheism and scientific (male) hubris (Hansen 1997, 578).

Comparing the stories of Prometheus (from Plato's *Protagoras* at least) and Frankenstein, one can see a significant difference on the level of narrative units. Called "*mythemes*" by Claude Levi-Strauss, "*functions*" by Vladimir Propp, and "*narremes*" by Algirdas Gremais and other prominent narratologists, these basic blocks break down myths, folktales, and other narratives into reusable modules that define characters and actions connected to them (cores and nuclei). Although the target of the rebellious action that both Prometheus and Frankenstein make is the same – Zeus and God respectively – the beneficiaries of the rebellion are completely different. In the Greek myth, Prometheus rebels against Zeus to bestow humankind with powers: the Titan steals fire to give it to humans. Dr Frankenstein, on the other hand, benefits no one other than himself with the gift of creation. He is not re-enacting the deeds of Prometheus for a betterment of the human race; instead, he imitates the very act of the gods' creation to spark to life the first specimen of the "new species". There is no intermediary between the gods and mankind. The human – or at least the science that Dr Frankenstein represents – takes the role of the Titan (Prometheus) and God (Zeus) in one single sweep. These two different types of rebellion toward the established seeds of power point to different possible types within semantic and mythical structures of Promethean stories. Panayot Karagyozov proposes situating them between opposing poles of two main motifs: theomachy and philanthropy (Karagyozov 2012, 96). The original Promethean myth gravitates towards philanthropy, and Mary Shelley's *Frankenstein* does so towards theomachy. Such a distinction functions well within classical studies up to the Romantic era. Later, however, in the post-Nietzschean world, with no God at the top of the pyramid of power relationships, Promethean motifs take different tones: in science fiction, the place of the gods and Titans is taken by totalitarian regimes, corporations, AI governments, and super intelligent aliens.

Promethean motifs in cyberpunk/biopunk games and fiction

In classical works of dystopian literature and science fiction of the near future, such as George Orwell's *1984* and Ray Bradbury's *Fahrenheit 451*, one can observe a significant disparity in the distribution of power, knowledge, and resources in society. Those with power are so omnipotent that any rebellion takes on titanic dimensions and often ends in failure. In cyberpunk, such disparities are softened. The characters of William Gibson's *Neuromancer* trilogy are hackers and programmers with a high knowledge of cyberspace – a virtual Wild West that can be accessed by skilful outlaws, even though it is not entirely a free domain and is under the control of corporations. Taking up the "cowboy myth" (Melzer 2019), hackers turn into "console cowboys" who are able to make a difference and bring the results of this difference to other disenfranchised parts of society.

The dog-eat-dog nature of the ultra-liberal capitalism of the cyberpunk genre – where fundamental inequalities and huge disparities between a rich minority and the rest of society are counterbalanced by a somewhat free-flowing distribution of cyberware, bioware, and military technologies – is a defining characteristic of the world of *Cyberpunk*. Created by Mike Pondsmith originally as a series of tabletop role-playing games (*Cyberpunk 2013*, 1988; *Cyberpunk 2020*, 1990), and recently made into the computer game *Cyberpunk 2077* (2020), the story takes place in Night City, which is a free city-state officially governed by local authorities yet overrun by gang wars and factually controlled by the Arasaka corporation. With ties to Yakuza mafia and dealing in military equipment, corporate security, manufacturing, cyber and biotechnologies, Arasaka constitutes a worldwide centre of power. Inevitably, most characters of the game will gravitate towards it with actions that range from collaboration to rebellion. As a result, Promethean themes present in the game are strongly Arasaka-oriented.

While the presence of biotechnology in the world of cyberpunk was not prominent in the futuristic world of *Cyberpunk 2013*, and in *Cyberpunk 2020* role-playing-games, in *Cyberpunk 2077* bioware unsurprisingly plays a major part in the main storyline. Whereas the first chapters of the Cyberpunk saga were indebted to the world of William Gibson's *Neuromancer* (with an emphasis on cyberspace technologies), the creators of *Cyberpunk 2077* had to account for some other major technological developments that happened between 2013 and 2077 in order to sustain a cohesive history of Night City and its surroundings. The shift from cyberspace to bioware is depicted in a series of books called "shards" that are scattered across Night City and are found by players and read via a direct port implanted in their heads. Somewhat surprisingly, genetic enhancement in 2077 is used mostly for crop, plant, and synthetic meat engineering. Commercially "gengineered" products are used in the beauty and sex industries, mostly for cosmetic reasons such as fluorescent tattoos.[69] In general, biotechnology in the world of *Cyberpunk* series mostly comprises technologies that integrate cyberware with the human body to enhance its capabilities and longevity and repair damage. This is

[69] Exceptions were present already in Cyberpunk 2020 and include a "Shukutei Biomed 'Mentor' Cerebral Enhancement" – a biocont that supplies the brain with hormones from a genetically engineered version of the pineal gland (normally active during the early stages of childhood), which results in a boost to one's intellectual abilities (https://www.cyberpunk2020.de).

confirmed in one of the shards: an introduction to *75 Years of Cyberware* by Tsutomu Takahashi:

> A century ago, losing a limb meant tragedy (…) Today, assuming the dismemberment victim is financially stable, loss of limb amounts to little more than a minor inconvenience. (…) With the advent of cyberware, employers in the second half of the twenty-first century have imposed requirements for skin, bone, muscle, organ and eye replacements in order to improve performance and workplace effectiveness. In extreme cases, security sector employees are commonly urged to undergo so-called full body conversions, or full cyborgization. (Takahashi, online)

The widespread use of cyberware, accelerated by making many enhancements a job requirement, had two major consequences, firstly in the structure of cybertech and biotech markets, and secondly in scientific development in these fields. The first made Night City abundant with ripperdocs, non-professional medical practitioners such as Victor Vector – a former boxer who can install a variety of cybernetic prostheses to anyone who can afford them. As a result of the second, a strong accessibility gap to cybertech – and especially biotech – is felt across the world of cyberpunk. Although a ripperdoc is legally allowed to install common cyberware, they are not allowed to supply a patient with a military grade implant. Even more so, experimental and breakthrough technologies are shrouded in secrecy, conducted under the auspices of the most powerful entities (Arasaka), and – when put on the market – they are accessible only to "the 1% who could be able to afford it" (Cyberpunk Wiki 2021). Such a distribution of resources establishes a truly Promethean setting: a precious technology (expensive advanced bioware) in the hands of the gods (Arasaka) that is "waiting" to be stolen by the Titans. The role of the latter is taken up by so-called "fixers", hired guns of Night City who negotiate their way between corporations, gangs, and the corrupt police force.

Promethean motifs in *Cyberpunk 2077* take a unique and ironic turn by introducing an element of split agency within the main protagonist. The central character of the game is a fixer called V who comes into possession of a revolutionary premium biochip called the Relic, which is able to preserve a copy of one's personality for future generations to interact with. After a failed heist in which V and her partner Jackie attempt to steal the updated version of the Relic from the heir of the Arasaka family, during which Jackie is killed and V almost dies, a test version of Relic 2.0 gets implanted in V's brain as the only way of keeping her alive after she sustains a shot to the head. It turns out that the Relic 2.0 was shipped with a personality construct of Johnny Silverhand, a punk rocker, rebel, and terrorist who had planted an atomic bomb under the Arasaka Tower back in 2023 and was killed shortly afterwards. It also transpires that if nothing is done, the

Relic will eventually kill V. Becoming inseparable, V and Johnny embark on a difficult, if not impossible, quest to extract the Relic while keeping V's body intact and preserving Johnny's construct. Because most advanced technologies are almost exclusively in the hands of the Arasaka mega-corporation, it is towards this adversary that V, Johnny, and the supporting characters direct their further actions. Depending on the players' choices, these actions are either forceful or cooperative and lead to different scenarios for ending the game. Astonishingly, the contradicting motifs and goals of V and Johnny remain stable across different endings. V wants things to go back to the state of affairs before the unexpected implanting of the Relic. Johnny, having woken up after more than fifty years and seeing Arasaka still in power, wants to resume his mission of destroying the corporation for the benefit of all, which is also a part of his own personal vendetta.

From the vintage point of basic narrative structures, the two different agendas of V and Johnny Silverhand represent two different types of stories; only one of them is closely aligned with the myth of Prometheus. In sections leading up to the failed heist and the implanting of the Relic, V's plot reflects the story of Frodo Baggins from J. R. R. Tolkien's *Lord of the Rings*. Just like Tolkien's Frodo, V comes into possession of a power she does not understand or own and which she wants to give back so that the world can return to a state of balance before the acquisition of this power. In stark contrast, Johnny does not want things to return to normal, but instead embarks on a mission to destroy or weaken Arasaka both in the real world and in cyberspace.[70] Despite V's frequent claims that Johnny's motivations are personal, egoistic, and driven by revenge, it is actually Silverhand (and not V) who embraces the philanthropic impulse on a scale of Promethean motivations ranging between theomachy and philanthropy. Weakening the leading source of power and making the advance technology more accessible to everyone represents the Promethean philanthropic motivation, whereas V's quest, driven solely by the will to survive, can be considered an individualistic theomachy with no intended impact on society. The paradox, irony, and perhaps lasting contribution of *Cyberpunk 2077* to the reservoir of modern adaptations of the myth of Prometheus is the narrative twist in which Promethean "mythemes" are not enacted by a Titan nor a mortal, but rather by a personality construct residing in a biochip, a form of AI built from the psychological traits of a dead person. Deeds of bravery are displayed by an agent who is already dead, and as such cannot be killed, whereas the human – V – cares mostly about herself and those close

[70] The cyberspace part of V's and Johnny's mission – in which they enter Mikoshi, a deep cyberspace, in order to find Johnny's former girlfriend and net runner Alt Cunningham – subscribes to yet another narrative structure reminiscent of Orpheus's journey to Hell in order to find Eurydice.

to her. Would V sacrifice her de-enhancement (getting rid of the biochip) for popular access to the experimental biotechnology that is killing her but which in the hands of ripperdocs could benefit the remaining 99% citizens of Night City? The game does not answer this question, perhaps to keep in line with the hedonistic and individualistic vision of Night City that Pondsmith created. Surprisingly, the motifs of Promethean sacrifice for better accessibility to science resurface in the ideas, assumptions, and actions of real-life characters involved in DIY science and biohacking.

The Windup Girl: A posthuman Prometheus

Cyberpunk 2077 demonstrated that identifying the sources of power towards which the Promethean rebellion is directed might be less enticing (at least in science fiction) than identifying rebellious individuals and their motifs. This is confirmed in *The Windup Girl* by Paolo Bacigalupi (2009). The story takes place in a post-oil future where international conglomerates accumulate power and resources and manage energy production by controlling agriculture and the genetic engineering of crops. World governments collapse under food and energy shortages created by such conglomerates, called “calorie companies”, whose weapon of choice is food patenting and genetically induced crop plagues. Thailand, where the story is placed, maintains its independence thanks to a secret genetic “seed bank” of crops and fruits that are plague-resistant. The Ministry of Trade and the Ministry of the Environment are two rival sources of power. The three significant characters of Bacigalupi’s novel are the main protagonist Anderson Lake, a representative of calorie companies on a secret mission to steal from the “seed bank”; Gibbons, a “gene ripper” working for conglomerates to create altered crops; and Emiko, a genetically engineered android geisha. Emiko represents the New People, a species of servants, soldiers, and workers created with the use of canine genes in order to be obedient, made sterile in order not to reproduce, and possessing a purposeful motoric dysfunction of stutter-stop motions in order to reveal themselves among humans as artificial and inferior “windups”.

The three main characters of *The Windup Girl* act in a setting that is ripe for resistance, being among corrupt government officials and greedy and ruthless representatives of corporate interests. They are also in a position of choice between obedience or disobedience, which also equips each of them with a Promethean attribute. Anderson Lake wants to acquire Thailand’s national treasure: the genetic seed bank. Potentially, he can choose to either fulfil his original mission and give the treasure (and power) to the corporation he works for, or he can give the seed bank to those who can make better use of it. Gibbons, a geneticist, has the power to alter crops and alter the New People to make them non-sterile and consequently start a new (posthuman) race that is resistant to crop related diseases and global environmental changes (Schmeink 2017, 115). Finally, Emiko can choose the obedience and servitude guaranteed by her design, or she can revolt against her "masters" who humiliate and torture her. This last choice potentially leads to the liberation of the New People and the start of their autonomous settlement outside of human populations. Each of the three protagonists is ready to put in motion the Promethean *"mythemes"* of rebellion, theomachy, and philanthropy in narrative actions related to acquiring power and granting it to those in need (Lake), sparking a new life (Gibbons), and starting a foundational period in the life of a new race (Emiko).

The narrative potential of *The Windup Girl*, which is dormant in pre-requisite actions and the position of protagonists along the narrative arch, is unleashed when Emiko murders a prominent minister and his entourage. The news about the homicide spreads and sparks chaos that leads to a popular uprising in Bangkok. At this moment, in a state of lawless flux before a new political order emerges, each of the characters can realize their narrative potential. Yet, once again, it is Emiko who acts. It turns out that only those with nothing to lose, the outlaws and those living on the margins of society, follow the rebellion phase with a delivery phase in which some positive resolutions occur. In this last instance, the narrative outcomes on the semantic level align with mythical units to form a Promethean connection. Emiko prepares herself and her New People for their imminent migration to some safe spaces outside Bangkok. Infected by one of the genetic crop diseases, Anderson Lake dies. Gibbons remains inactive, although his words suggest that he is ready to take the role of the ultimate helper, a Promethean persona who – through the gift of fertility – creates a new human race out of the New People. As Gibbons addresses Emiko:

> Nothing about you is inevitable […] someday, perhaps, all people will be New People and you will look back on us as we now look back at the poor Neanderthals […] you cannot be changed, but your children – in genetic terms, if not physical ones – they can be made fertile, a part of the natural world.

Biopunk scholars such as Lars Schmeink and Heather I. Sullivan (2012, 522–523) agree that Bacigalupi's novel opens a possibility of a "truly posthuman future and the eventual replacement of the human" (Schmeink 2017, 115–116). Both *Cyberpunk 2077* and *The Windup Girl* suggest that it might not be humans who initiate such a future. The Prometheus of the posthuman era, these works suggest, will already be a posthuman!

The biohackers of CRISPR

A system of second-order meaning which lets us identify a structure of relations that either invoke or accommodate a myth (Barthes 1991, 113–114) is at play in literary fiction and computer games as well as in any conversation: in the words, actions, and motivations of its participants. If it relates to areas not fully explored (such as genetic engineering) and to things not yet fully known (such as the consequences of gene therapy), categories and distinctions of a given discourse can be strongly indebted in myth. It is within those unexplored territories that myths serve as a source through which "culture gives meaning to behaviour" (Culler 2001, 26) and myths reveal their main function of establishing models for social behaviour (Eliade 1963, 137). When facing uncertainties, people draw from a reservoir of cultural imagination supplied by popular science, science fiction, and speculative fiction, blurring the boundaries between real life and the fictional world even further.

In 2018 the biggest controversy to date shook the world of CRISPR-cas9 genetic engineering when Chinese scientist He Jiankui edited human embryos in two twins, violating at least ten internationally established bioethical rules (Krimsky 2019, 19). Just months later, Netflix streaming platform presented *Unnatural Selection*, a TV documentary series that gave voice to multiple parties involved in current biotechnology such as scientists, corporations, patient protection organizations, and – most prominently – biohackers. Although the controversy of Jiankui is mentioned only in the last episode of the documentary, the public reaction to his bioethical violations presents an illustrative context for the discussion on Netflix. Notwithstanding the response from the scientific world, where Jiankui was quickly labelled as a "rouge scientist", in their effort to illustrate the weight of his actions, the public media reached for proven metaphors and shortcuts, calling the Chinese biologist "Frankenstein" and a "mad genius" (Low 2018), directly evoking the Promethean myth and its Romantic and later cultural renditions.

Through a diverse cast of characters, *Unnatural Selection* introduced viewers to different perspectives on gene therapy, its accessibility, and ethical limits. The opinions of geneticists, biologists, bioethicists, ecologists, patients, and their families were voiced. Divergent groups of interests were represented, with one side of the spectrum occupied by official medical institutions, bioscience research centres, and corporations that try to make gene therapy commercially available.[71] The other side is represented by biohackers, patients, and families who demand the wider accessibility of treatment.[72] The locations range from dog kennels and the garages of biohackers to designer baby clinics and the United States Congress. The discussions take place against the dramatic background of individual patients with life-threatening genetic disorders, whole populations with an endemic problem (such as malaria), and whole ecosystems on the brink of collapse (e.g. because of an overpopulation of rodents) and in desperate need of the solutions that genetic engineering can potentially deliver.

Because the subject of the documentary is technology in a nascent stage, and the application and accessibility of new gene therapies on people in general, the documentary abounds in motifs and patterns of behaviours associated with the Promethean myth. To trace these *"mythemes"* in the actions and motivations of protagonists who "trigger" the Promethean associations, a closer look is necessary at the underlying grammar upon which the mythical and narrative patterns are built. As I have demonstrated during the discussion on *Cyberpunk 2077*, *Frankenstein*, and *Windup Girl*, one can assess if a narrative aligns with a myth by identifying similar narrative patterns built by few basic components: an agent, its motivation, and its action. A further comparative study is possible by pointing out the vectors of actions and motivations, for example, by identifying the main beneficiary and the main adversary of agents' actions. The second step of such procedure moves beyond linguistic "actantial systems" of narrative analysis proposed by pioneering narratologists such as Algirdas Gremais and Claude Levi-Strauss (Rosenbaum 2019, 5) towards a more nuanced and psychological area of personal traits and motivations. In other words, we can see several Prometheuses in *Unnatural Selection* by examining their actions (what they do) as well as by embracing their personal motivations and ethical horizons (who they are).[73]

[71] These groups of interest were represented by Jennifer Doudna (an American biochemist and Nobel laureate for CRISPR), Victor Dzau (President, United States National Academy of Medicine), Jeffrey Kahn (a professor of bioethics), Preston Estep (the CSO of Veritas Genetics), and Katherine A. High (the CSO of Spark Therapeutics) among others.

[72] The most notable include David Mitchell, the head of Patients for Affordable Drugs; Aaron Traywick, the late life extension activist and the CEO of Ascendance Biomedical; Josiah Zayner, the biohacker and artist; and Tristian Roberts, an HIV patient advocating for self-therapy.

[73] A comprehensive methodology suitable for discerning structures and patterns out of any story, applicable, for example, in an analysis of computer games, was proposed by Richard Rosenbaum as part of his updated theory of "narremes": a basic (non-narrative) unit of a narrative composed of single states of the represented world (actant, locus) and their values. It is detailed and precise. Most of Rosenbaum's article is devoted to an exemplary analysis of a simple nursery rhyme. A comparative study of "narremes" in a computer game, two science-fiction books, and one lengthy documentary would take up even more space. In this article – whose subject is not solely focused on the semiology of narrative

In the myth "speech" that *Unnatural Selection* delivers alongside its cinematic and linguistic utterances, the most obvious candidate for the role of Prometheus on the mythical level is Jennifer Doudna, one of the inventors of the CRISPR genome editing tool: an invention that took the world of bioscience by storm, sparked hopes in patients with genetic diseases, and hugely accelerated the development of the biohacker scene. We should consider Doudna's main adversary to be genetic diseases, or more generally the faults and errors of evolution that cause genetic disorders. CRISPR is a tool to remedy these faults, and the main beneficiary is humanity in general and particularly science. Throughout the documentary, which uses footage of the public appearances of the American biochemist in political forums and international news outlets, it becomes clear that ethical considerations play a fundamental part in Doudna's message to society. As a signatory to a proposal for a worldwide moratorium on any clinical application of germline gene therapy (Nguyen 2019), Doudna advises patience, control, and scientific rigour concerning access to and the implementation of CRISPR technology. On the mythical level, the initiator of the "CRISPR revolution" is a cautious Prometheus: someone who uncovers the secret (of evolution) and delivers the technology but does not want it to go unchecked.

The same Promethean pattern of delivering new technology to people with a basic motivation to benefit them is played out in a strikingly different way by Josiah Zayner, an American biohacker, artist, and former NASA scientist. *Unnatural Selection* introduces Zayner in quite a Promethean setting: it is early in the morning, and he is in his own garage in Palo Alto packing gene modification toolkits into cardboard boxes for pick-up and delivery to customers interested in DIY science. As a declared biohacker, Zayner wants the CRISPR technology to be accessible for everyone now and without the need to "wait in line" for "Big Pharma" or "Big Science" to decide when this should be done. Apart from creating a distribution network for DIY gene editing and publicly advocating for accessibility of the tools, Zayner delivers performance-like statements to achieve this goal. During one of them, he publicly injected himself with CRISPR-modified DNA for enhanced muscles. The mythical "speech" renders Zayner as a rebel Prometheus, someone who steals the tools from the gods to hand them out to everyone for the self-cure of self-enhancement. His main adversary is the bioscience establishment, which makes gene therapy expensive, inaccessible, and highly controlled through patent procedures and practices. Although he is neither a rouge

– Rosenbaum's theory of the "narreme" is only referred to and not fully applied. Future studies of mythical motifs in contemporary discourse might find the renewed theory of the "narreme" to be highly beneficial.

scientist nor a Frankenstein who creates a monster in his garage – Zayner's CRISPR toolkits are based on experiments on animals and bacteria – the Promethean alter-ego of the American biohacker can be considered both a trickster and a thief: an archetype attributed to Aristophanes's retelling of the myth rather than Plato's (Priestman 2018, 47). And yet his actions should not entirely be labelled as theomachy. Zayner's activism and dedication to the cause to the accessibility of genetic tools also bear the traits of Promethean philanthropy!

The two examples of Doudna and Zayner voice the interests, ambitions, and considerations of two prominent groups that *Unnatural Selection* sets against each other in the film. Yet they are just the tip of the iceberg. Other people associated with CRISPR technology, and those who self-associate with it, bear Promethean traits or aspire to bear them just as visibly. A telling example is Tristan Roberts, an HIV patient and biohacker who injects himself with an untested anti-HIV genetic treatment with the intention of rapidly testing the effectiveness of treatment and potentially helping other HIV patients with a proven cure. Of course, the expressed intentions do not necessarily align with the implied and hidden ones. Using language, narrative patterns, and the imagery of myths (in this case, the myth of Prometheus) can help in clarifying divergent intentions and goals across the public discourse on the advancement of biotechnology. A useful starting point might be a basic exercise of identifying main actors within the given discourse: the gods, the Titans, and the beneficiaries and adversaries within a story. In our case, a Titan is someone such as Doudna; the gods – not known by their names – are the abstract yet deterministic forces of evolution. Importantly, however, when one shifts the perspective and looks at the same technology from a different point of view, Doudna functions as a god, and the role of the Titans who steal from the gods is taken by the biohackers. In the same manner, one can analyse discussions surrounding the highly controversial case of Jiankui and his modified twin patients. What kind of Prometheus does Jiankui want people to believe he is? And what kind does he actually turn out to be? I hope that myths, with their strong, clear, and simplistic structures can help us discern these nuances and find clear answers to such questions.

Blurred boundaries between fact and fiction

In terms of the accessibility of genetic modification, the future painted in *Unnatural Selection* can be darker than the scenarios presented in dystopian science fiction. In the cyberpunk world of rampant capitalism and post-government order, where the rules of law are written by corporations and street gangs, access to basic gene therapies and bio-enhancements is nonetheless easier than in today's United States. Gene therapy that would prevent blindness in a child is worth almost one million dollars. Although this only means that our present becomes a future once described by authors of fiction, it also indicates that within the discourse on the social and economic implications of biotechnology, a literary vision might have a similar weight as an informed opinion. If not, then it can at least function as a convenient shortcut in communicating complex ideas in a few words with a reference to a commonly known cultural artefact.

The sensation of blurred boundaries between real life and fiction is confirmed by David Mitchell as the head of the Patients for Affordable Drugs organization, who in *Unnatural Selection* compares the accessibility situation in the United States to Ridley Scott's movie *Blade Runner* (1982), where corporations are in control "of the way the world business is conducted and our lives are lived". In another place, Jeffrey Kahn, a professor of bioethics, draws a comparison to the film *Gattaca* by Andrew Niccol (1997) when referring to accessibility and the social disparities it might bring.

It is in this grey area on the border of reality and fiction that Jeffrey Steinberg from the Fertility Institute makes his prediction about the near future of designer babies and our approach to the issue of choosing selective DNA traits for our own children. Steinberg, who was involved in the first successful in vitro fertilization in the United Kingdom and who currently advocates pre-implantation genetic diagnosis (PGD) (Naik 2018, 393), stated in the documentary that, whether we want it or not, designer babies are the future, just as forty years ago the future was represented by IVF. As the British IVF specialist pointed out, "In 100 years, all of us will be designer babies." If this happens, current ethical dilemmas about the use of genetic therapies on humans and about the limits that need to be imposed might become suspended and even obsolete. In this scenario, reality will confirm intuitions of science fiction authors who, just like Bacigalupi in *The Windup Girl*, predict that our posthuman future will be decided not by humans but posthumans. In other words, it will be edited humans – designer babies – that will set the tone for the future discussion on the limits of bioscience's interventions into the human genome. If so, the future may not look entirely the way the inventors of CRISPR had envisioned.

Conclusion

Stories of human hubris pushing us too far, a utopian promise of eradicating diseases and imbalances in environment, and pushing the boundaries of science outside its traditional environment are all common themes of science fiction books and games that relate to scientific development towards human enhancement which have a strong foundational core rooted in myth. The ancient Greeks devised two myths that accommodated these themes. The first myth is that of Prometheus. It is a story of origins and rebellion against the gods that successfully benefits humanity with the "powers" of fire and toolmaking. The second myth is that of Deadulus and Icarus. It describes later stages in human development when the gifts stolen from the gods by Prometheus result in bold scientific inventions, especially flying. Ever since Mary Shelley's *Frankenstein*, G. H. Wells's *The Island of Dr Moreau*, and J. B. S. Haldane's influential essays on the future of science, the myth of Prometheus has accompanied human endeavours into mechanical, cybernetic, and biological enhancement.

This comparative study's goal was to identify appearances of Promethean *"mythemes"* – of rebellion and gifting humanity with "powers" on a scale marked by theomachy and philanthropy – within the contemporary discourse on biotechnology, where the imaginary worlds of fiction and real live developments merge into a diverse yet cohesive range of voices about our future.

In the analysed fictional examples, the mythic perspective was able to uncover some interesting dynamics between the motivations of the *Cyberpunk 2077* protagonists – V and Johnny Silverhand. The latter – a cybernetic entity and mental construct extracted from the memories of a deceased rebel punk rocker – turned out to be much more uncompromising and more "philanthropically" oriented towards humanity than the main character V, which is something that goes against the gamers' and reviewers' perceptions of the two heroes. This seemingly odd finding that someone who is posthuman (and, in this case, also posthumous) can be more benevolent and future-oriented than a human is confirmed in the suggestive ending of *The Windup Girl*, where it is also up to a posthuman to establish a better life for herself and the environment.

The subject of posthumans penetrates the entire discussion in the documentary *Unnatural Selection*. Despite a whole range of colourful Promethean figures (who may well be genuine and self-styled and highly surpass their fictional counterparts), the main take-away from the mythical reading method is the discovery of a major shift in distributing the godly "powers" of genome editing. Namely, gene editing technology may be directed toward benefiting not just a human but (already) a posthuman. This is a shift that may be yet to come or that might have already happened. Steinberg's somewhat sarcastic remark that in one hundred years the controversy of designer babies will be no longer controversial because most people will be designed can become a self-fulfilling prophesy. Although the moral considerations expressed by Doudna and Kahn are well founded and necessary today, the lines drawn by these considerations can be abolished in the future. Indeed, if Steinberg's clinics proliferate and prosper globally, in one hundred years there might be not many new-born humans anymore. There will be (slightly modified) humans who will write the rules of their own posthuman genome editing.

Acknowledgement

This work was supported by Slovak Research and Development Agency contract No. APVV-17-0064.

Bibliography

Arieti, James A., and Roger M. Barrus (eds.). 2010. *Plato's Protagoras*. Lanham: Rowman & Littlefield Publishers.

Barthes, Roland. [1957] 1991. *Mythologies*. Trans. by Annette Lavers. New York: Noonday Press.

Bourdieu, Pierre. 2005. *The Social Structures of the Economy*. Trans. by Chris Turner. Cambridge: Polity Press.

Breen, Myles, and Farel Corcoran. 1982. "The Myth in the Discourse." A paper presented at the Annual Meeting of the Central States Speech Association (Milwaukee, WI 15–17 April). Accessed April 20, 2021. https://files.eric.ed.gov/fulltext/ED213053.pdf.

Cardozo, Karen, and Banu Subramaniam. 2008. "Genes, Genera, and Genres: The NatureCulture of BioFiction in Ruth Ozeki's All Over Creation." In *Tactical Biopolitics: Art, Activism, and Technoscience*, ed. by Beatriz da Costa and Kavita Philip, 269–287. Cambridge, MA: The MIT Press.

Culler, Jonathan. 2001. *Barthes: A Very Short Introduction*. Oxford: Oxford University Press.

Eliade, Mircea. 1963. *Myth and Reality*. Trans. by Willard R. Trask. New York: Harper & Row.

Foucault, Michel. 2019. *Ethics: Subjectivity and Truth. Essential Works of Michel Foucault 1954–1984*. Trans. by Robert Hurley. London: Penguin.

Freud, Sigmund. 2010. *The Interpretation of Dreams*. Trans. by James Strachey. New York: Basic Books.

Gomel, Elana. 2011. "Science (Fiction) and Posthuman Ethics: Redefining the Human." *The European Legacy* 16, 3: 339–354. DOI: https://doi.org/10.1080/10848770.2011.575597.

Haldane, John Burdon Sanderson. 1995. "Daedalus; or, science and the Future." In *Haldane's Daedalus Revisited*, ed. by Krishna R. Dronamraju, 23–51. Oxford: Oxford University Press.

Hammond, Emma. 2018. "Alex Garland's Ex Machina or the Modern Epimetheus." In *Frankenstein and Its Classics: Modern Prometheus from Antiquity to Science Fiction*, ed. by Jesse Weiner, Benjamin Eldon Stevens, and Brett M. Rogers, 190–205. London, New York: Bloomsbury Academic.

Hansen, Mark. 1997. "'Not Thus, After All, Would Life be Given': Technesis, Technology, and the Parody of Romantic Poetics in Frankenstein." *Studies in Romanticism* 36: 575–609.

Karagyozov, Panayot. 2012. "Prometheism Degenerated: On Material from Ancient Greek and Polish Literature." *The Polish Review* 57, 1: 95–121.

Krimsky, Sheldon. 2019. "Ten Ways in Which He Jiankui Violated Ethics." *Nature Biotechnology* 37: 19–20. DOI: https://doi.org/10.1038/nbt.4337.

Lazaratto, Maurizio. 2002. "From Biopower to Biopolitics." *PhilPapers* 13: 112–125.

Lewontin, Richard. 1996. *Biology as Ideology: The Doctrine of DNA*. Toronto: House of Anansi Press.

Low, Zoe. 2018. "China's Gene Editing Frankenstein He Jiankui, dubbed 'Mad Genius' by Colleagues, had Early Dreams of Becoming Chinese Einstein." *South China Morning Post*, November 27. Accessed June 15, 2021. https://www.scmp.com/news/china/society/article/2175267/chinas-gene-editing-frankenstein-dubbed-mad-genius-colleagues-had.

Mayor, Adrienne. 2018. *Gods and Robots: Myth, Machines, and Ancient Dreams of Technology*. Princeton: Princeton University Press.

Melzer, Patricia. 2019. "Cyborg Feminism." In *The Routledge Companion to Cyberpunk Culture*, ed. by Anna McFarlane, Lars Schmeink, and Graham Murphy, 291–299. London: Routledge.

Morales, Salomé S. 2013. "Myth and the Construction of Meaning in Mediated Culture." *KOME: An International Journal of Pure Communication Inquiry* 1, 2: 33–43.

Naik, Gautam. 2009. "A Baby, Please. Blond, Freckles – Hold the Colic." *The Wall Street Journal*, February 12. Accessed June 6, 2021. https://www.wsj.com/articles/SB123439771603075099.

Nguyen, Tim. "Pioneers, Unparalleled Promises, and the Moratorium." *Stemosphere*. Accessed April 12, 2021. https://stem-o-sphere.org/pioneers-unparalleled-promises-and-the-moratorium/

Patterson, Meredith L. 2010. "A Biopunk Manifesto." Accessed April 12, 2021. https://maradydd.livejournal.com/496085.html.

Pentecost, Claire. 2008. "Outfitting the Laboratory of the Symbolic: Toward a Critical Inventory of Bioart." In *Tactical Biopolitics: Art, Activism and Technoscience*, ed. by Beatriz da Costa and Kavita Philip, 110–24. Cambridge, MA: The MIT Press.

Priestman, Martin. 2018. "Prometheus and Dr Darwin's Vermicelli: Another Stir to the Frankenstein Broth." In *Frankenstein and Its Classics: The Modern Prometheus from Antiquity to Science Fiction*, ed. by Jesse Weiner, Benjamin Eldon Stevens, and Brett M. Rogers, 42–58. London, New York: Bloomsbury Academic.

Rogers, Brett M. 2018. "The Postmodern Prometheus and Posthuman Reproductions in Science Fiction." In *Frankenstein and Its Classics: The Modern Prometheus from Antiquity to Science Fiction*, ed. by Jesse Weiner, Benjamin Eldon Stevens, and Brett M. Rogers, 206–227. London, New York: Bloomsbury Academic.

Rosenbaum, Richard. 2019. "Toward a Renewed Theory of the Narreme." *The American Journal of Semiotics* 35, 1/2: 187–215. DOI: https://doi.org/10.5840/ajs201982255.

Schmeink, Lars. 2017. *Biopunk Dystopias Genetic Engineering, Society, and Science Fiction*. Oxford: Oxford University Press.

Shelley, Mary Wollstonecraft. [1818] 1999. *Frankenstein, or the New Prometheus*. London: Wordsworth.

Stockstill, Ellen J. 2019. "Review of the work: Frankenstein and Its Classics. The Modern Prometheus from Antiquity to Science Fiction, ed. by Jesse Weiner, Benjamin E. Stevens, and Brett M. Rogers." *Sciences Fiction Studies* 46, 3: 659–662.

Sullivan, Heather I. 2012. "Dirt Theory and Material Ecocriticism." *ISLE: Interdisciplinary Studies in Literature and Environment* 19, 3: 515–531. https://doi.org/10.1093/isle/iss067.

Takahashi, Tsutomu. *75 Years of Cyberware*. Accessed August 10, 2021. https://cyberpunk.fandom.com/wiki/75_Years_of_Cyberware,_by_Tsutomu_Takahashi.

Thacker, Strom C., and John Gerring. 2008. *A Centripetal Theory of Democratic Governance*. Cambridge: Cambridge University Press.

Weiner, Jesse, Benjamin Eldon Stevens, and Brett M. Rogers (eds.). 2018. *Frankenstein and Its Classics: The Modern Prometheus from Antiquity to Science Fiction*. London, New York: Bloomsbury Academic.

Chapter 7
Expanding the Boundaries of Literature as an Interdisciplinary Plot in Art-Science

Bogumiła Suwara

Abstract: This chapter focuses on an analysis of interdisciplinarity. The main objective is to present digital literature and literary work as an interface between the two autonomous systems and disciplines of literary science and artificial language. By examining digital literature and code poetry as an interdisciplinary product, the present author shows that the phenomenon of linguistic creation can be defined by the same term in the two fields even though the meaning itself is not identical. This can be demonstrated through the prism of a hermeneutic analysis of the text as it is based on different activities in different disciplines, thus blurring the boundaries. However, the blurring of boundaries does not lead to an integration of knowledge but rather to the establishment of the new subdiscipline of code studies. It also takes the form of a dispute between those views which are typical for the field of literary theory and those of the programming community. In this chapter, the dispute is analyzed upon the basis of the typology of interdisciplinarity by Andrew Barry and Georgina Born. In the case of art-science, they propose conceiving interdisciplinarity through the prism of the three pillars and logics of accountability, innovation, and ontology.
Keywords: Interdisciplinarity, the blurring of boundaries, the integration of research, code poetry, the interface of literature.

Introduction

Despite the fact that the beginnings of digital literature were accompanied by the perception of an insurmountable difference between the culture of the Gutenberg Galaxy and theoretically presented revolutionary changes determined by Web 2.0 information technology (Landow 1992), digital literary works have become an established part of academic discourse. Digital literature is also a part of the process of academic education (Brillenburg Wurth 2017) and research (Aarseth 1997; Hayles 2002; Ryan and Thon 2014). Digital literature has frequently been discussed in perspectives concerning new media, media communication, remediation (Bolter 2011; Bolter and Grusin 1999), media-specific analysis (Hayles 2002), archaeology, media variability (Zielinski 2013), intermediality (Higgins 2000; Heimej 2014), and the technologizing of the word (Ong 1977). Researchers who are sensitive to the co-evolution of technology and society have emphasized the correlation of digital literature with the attributes of the information network society, media society, and post-media society (Castells 2008; Barney 2008; de Kerckhove 2001; Schmidt 2010; Rankov 2006; Kluszczyński 2001). These contexts suggest that academic reflections on digital literature do not take place within a single discipline. The interdisciplinarity aspect of digital literature has recently been highlighted by Andrew Barry and Georgina Born in *Interdisciplinarity: Reconfigurations of the Social and Natural Sciences* (2013). Based on ethnographic research of art-science institutions, practices, and administrators in the United Kingdom, United States, and Australia (CRESC 2014), "new media" and "digital art" was included within the framework of "art-science", which was defined as follows:

> We propose that art-science should be understood as a multiplicity, and that part of its interest lies in not being reducible to the imperative to render scientific knowledge more accessible or accountable. Indeed art-science poses definitional and conceptual challenges since, while it exists as a practical, intentional category for artists and scientists, cultural institutions and funding bodies, it forms part of a larger heterogeneous space of overlapping interdisciplinary practices at the intersection of the arts, sciences, and technologies. This includes new media art and digital art, interactive art and immersive art, and bio art and wet art (Wilson 2002; Da Costa and Philip 2008; Leonardo 2012), while these domains about adjacent interdisciplinary scientific and technological fields from robotics, informatics, and artificial and embodied intelligence to tissue engineering and systems biology. There is thus a ferment of activity but little codification: "art-science" amounts to a pool of shifting practices and categories that are themselves relational and in formation. (Barry and Born 2013, 248)

In this understanding, art-science summarizes previous opinions formed upon the basis of analyses of the bilateral relations between science and art, including opinions of undervalued authors of new media art who were asked by scientists to create visualizations and video presentations of research results (Vesna 2011). From the artists' perspective, it was only a matter of their skills in using video technologies. Both parties were aware that there was no crossing of disciplinary boundaries as the artists did not participate in obtaining the scientific findings. And yet artists are often convinced about the benefits they can bring to the processes of scientific experimentation through their inventive and imaginative abilities. As one study suggested, such optimism is rarely shared by scientists (Groth, Kääriäinen, Pevere, and Niinimäki 2019, 2020).

When investigating the success of scientific projects carried out in the presence of artists, researchers have come to a conclusion confirmed by both artists and scientists. A scientific art project can only be successful if both parties address a common problem from the perspectives of both disciplines. To level the playing field, a change of positions at the subsidy granting process is suggested, whereby artists ought to apply and administer funding on behalf of both parties (the concept of collaborative sustainability as a dominant attribute of interdisciplinarity [see

Miller 2001]; or the equality of opportunities). Barry and Born have highlighted the difficult situation in terms of financial uncertainty and the lack of the successful implementation of interdisciplinary research aimed at linking science and the arts into the form of tertiary institutions and processes of academic education. Interdisciplinary research (such in art-science) is presented by Barry and Born in accordance with the historical ideas of the predecessors of transdisciplinarity and interdisciplinarity.[74] It is necessary to familiarize society with scientific knowledge, especially in situations where technology is intensively affecting its development. The aim of interdisciplinarity should therefore be the mediation of knowledge. Supporters of the transdisciplinarity movement in Europe attribute an important societal significance to obtaining scientific and other findings by, for example, involving the wider public in the processes of testing, evaluating, and implementing knowledge simultaneously in different disciplines (Miller 2021). Moreover, it has been necessary to find a solution to eliminate the negative effects of the splitting of university disciplines and the dispersion of scientific findings and knowledge. The initiators of interdisciplinarity considered the search for methods and processes to link otherwise disparate results into holistic units as a very promising solution. This was demonstrated by the interdisciplinarity researcher Raymond C. Miller on multiple levels. Indeed, "Interdisciplinarity is an analytically reflective study of the methodological, theoretical, and institutional implications of implementing interdisciplinary approaches to teaching and research" (Miller 2021).

In the case of art-science, Barry and Born propose looking at interdisciplinarity through the prism of the three pillars and logics of accountability, innovation, and ontology:

[74] Several authors link the terms to their first professional use in a 1972 OECD report entitled "Interdisciplinarity: Problems of Teaching and Research in Universities" (Apostel 1972; Klein 1990; Barry and Born 2013; Miller 2010). From a historical perspective, there was a belief held by scholars that scientific knowledge was being devalued and inhibited by the division (and fragmentation) of academic disciplines. They saw a remedy to this condition in the integration of knowledge and findings that would be achieved through unifying research schemes and academic education (e.g. general systems, Marxism, and structuralism). Interdisciplinarity was the vision for the direction of academic inquiry and education. In these contexts, "transdisciplinarity" was promoted by some researchers as a set of specific propositions for achieving the desired integration of concepts and methods. Prominent promoters of transdisciplinary practices were associated with France and Germany (Jantsch 1972; Nicolescu 1985), while the practice of running interdisciplinary universities was more dominant in the United States and the United Kingdom. This is probably why the terms are sometimes presented as synonymous, as the use of one particular variant was determined geographically (Toomey, Markusson, Adams, and Brocket 2015). Certainly, much work has been produced over the last fifty years in attempting to determine the differences and similarities between transdisciplinarity and interdisciplinarity, particularly in relation to specific disciplines and dominant social issues (e.g. Nowotny 2001; Stock and Burton 2011; Yetiv and James 2017).

> By the logic of accountability, we refer to a series of ways in which scientific research is increasingly required to make itself accountable to society. By the logic of innovation, we draw attention to a range of arguments about the need for scientific research to fuel industrial or technological innovation and economic growth – a discourse that, while it has a long history, has exhibited a particular intensity in recent decades, [...] logic of ontology: an orientation in interdisciplinary practices towards effecting ontological transformation in both the object(s) of research and relations between the subjects and objects of research. (Barry and Born 2013, 248–249)

Barry and Born emphasize three logics of interdisciplinarity, but they are also aware that they have arrived at a typological generalization based on empirical findings and insights.[75] They therefore acknowledge that the three logics have not been firmly defined and have not provided a general analytical process in the field of art-science.[76]

This chapter works with the idea of interdisciplinarity in order to present digital literature and literary works as an interface of two autonomous systems and as an interface of literary science and artificial language. This is an interface of two disciplines. Research into digital literature and code poetry as an interdisciplinary product shows that the phenomenon of linguistic creation can be named by the same term in two fields, even though the meaning is not identical. Furthermore, the same approach, such as the hermeneutic analysis of a text, is based on different activities in different disciplines and blurs the boundaries between them. As a consequence, instead of synthesizing knowledge, new subdisciplines (such as code studies) are created. This chapter will (discretely) correlate the collected findings on the conflict between views legitimized in the field of literary theory and views from the programming community with the typology of interdisciplinarity by Barry and Born (2013) and Barry, Born, and Weszkalnys (2008).

[75] Advocates of interdisciplinarity are no longer inclined to seek universal methods or approaches to the interdisciplinary acquisition of knowledge and interdisciplinary inquiry. Trying to synthesize the findings of different disciplines has proven unrealistic. Instead, it seems that the interdisciplinary investigation of a particular problem can proceed as a dispute or disagreement. The consensus is that interdisciplinary inquiry often uses systems theories, information theories, data, and diverse concepts. Currently, the concept of sustainability is a predominant issue. A balanced view of the idea of interdisciplinarity can be found in "The National Academies Report" (2005). The report states that "there are four 'drivers' for interdisciplinary research: the inherent complexity of nature and society, the need to explore areas that are not confined to a single discipline, the need to solve societal problems, and the power of new technologies" (2005, 40).

[76] "The three logics of interdisciplinarity, then, have a different prominence and distribution, and are differently entangled, in the sites of art-science that we researched" (Barry and Born 2013, 249).

Digital literature on the path from a monodisciplinary approach to interdisciplinarity (art-science)

When searching for arguments for expanding towards electronic literature, the proposal by a librarian and curator of art book exhibitions (Museum of Modern Art, New York) to systematize these "book publications" – or interfaces developed between the book publication system and the visual arts – serves as a prototype. Apart from the category of the ordinary book ("just books"), Clive Phillopot singled out the category of the book as a work of art ("bookworks") and the book as an object ("book objects"). The latter only refers to the idea of the "book" (e.g. its structure and the idea of a book in a certain period). In a less formal sense, this reference can be materialized by an art installation as a symbolic or metaphorical reference to a particular publication (Rybson 2000; Tribe 2009, 54). At the same time, this systemization, which is inclined towards alternative solutions, has helped to make the shift from the closed and autonomous system of the printed book towards interactions with different systems such as electronic books and electronic literature.

It is a historical fact that producers, promoters, and scholars of electronic literature have made efforts to conceptualize the book as a separate and autonomous system where the rules of paper literature are not essential or do not apply. This attitude is understandable given the genealogy of digital literature. One of the primary motivations was to test the limits and possibilities of software applications for non-professional purposes within academic education – for example, for an experience of a liberated invention that at times provided an aesthetic experience or an opportunity to acquire aesthetic value. In this context, the most often mentioned hypertext projects are *Afternoon* by Michael Joyce (1987), *Patchwork Girl* by Shelley Jackson (hypertext collage, 1995), *Sunshine '69* by Robert Arellano (interactive web novel, 1996), and *Sintext* by Pedro Barbosa (1992 and 1996). The same trajectory of inventive use was continued by the producers of popular PowerPoint presentations at the turn of the millennium which mimicked a computer game, created a short story, or multiplied the media dimensions of set design and the narrative components of theatrical performances (Suwara 2012). The consequence of this was not only the sheer inventiveness of using a particular application but also the shift in technological skills towards an online public (a demand of the transdisciplinarity movement). The enthusiasm of producers and consumers for new digital formats of computer-based communication has sometimes influenced digital literary criticism's ideas of genre formats.

As a result of the pressures of the tendency for interdisciplinary in literary science and criticism, the notion that the media formats used in digital works should be introduced into literary critical discourse has garnered some legitimacy. For instance, criticism of a literary hypertext should be presented only in hypertext format – that is, in HTML or Storyspace. Examining computer games primarily through active players is a similar and somewhat more justified claim (Taylor 2006).

The programming community (which is still the dominant community amongst producers and consumers of electronic literature) promoted and spread the expectation that discourses concerning electronic literature and programmable media would become dominated by the language used by programmers themselves. Furthermore, electronic works would exclusively use terms and concepts that were established and used in the field of programming. The result of this tendency has resulted in a large presence of new terms in such discourses. To some extent, this has been perceived as exaggerated and unnecessary from the creators of academic reflections on literature (Hejmej 2013, 120).

There was certainly at least one rational reason for the tendency to autonomize electronic literature research – namely, the need for digital literacy among scholars and critics of electronic literature. According to some scholars, this reason is the result of the correlation between the development of information and communication technologies and the changes they have brought about in society. In other words, the information, digital, and media revolutions first interfered with the functioning of the information society, then the media society, and then the post-media society. However, digital literacy has predicted and caused a tension that can be interpreted as a "dispute" between programmers' claims to think of electronic literature as a self-created technological product and arguments to examine digital literature through the prism of the tools of literary science. This has been developed by generations of literary scholars who have sought to cover new subjects ("objects") with the umbrella of literary science and literary communication (e.g. the categories of author, reader, text, and intertextuality). This dispute clashed with the concept of hermeneutics, which was developed mainly for book texts created in a natural language, the concept of the ontology of literary works based upon phenomenology and structuralism (as opposed to the processual concepts of ontology of works of art), and the concept of the "unsaid" in art (Silverman 2014).

At the initial stage of the dispute, works by Manuel Castells, and especially George P. Landow (Hyper/Text/Theory, Hypertext, and Hypertext 2.0), published from the mid-1990s, dealt with the first prototypes of electronic literature in the form of hypertext. At a discursive and theoretical level, this stage promised far greater possibilities for hypertext than what was actually delivered (Pang 1998). A quasi-consensus between the potential of hypertext and the theory of text and writing was achieved by Landow using terminology taken from Roland Barthes ("link", "node", and "hypertext") and concepts introduced in the works of Gilles Deleuze and Félix Guattari ("rhizomatic structure") as well as Jacques Derrida ("the effect of inscription"). It seemed that the notion of technological determination and the genealogy of digital projects would be a fundamental basis for investigation and reflection. A slightly less definite view was presented by the programmer Mark Bernstein. Although he generalized empirical experience with the structuring of hypertexts and developed a detailed typology of literary hypertexts (Bernstein 2002), Bernstein was aware that this was a direction that may or may not make much sense to pursue. (Bernstein continues to support the creation and publication of literary hypertexts.)

The logic of accountability

Another aspect of the dispute over the definition of the disciplinary boundary of academic research into electronic literature can be legitimately linked to the open source software community (Šoka 2011). This is the acceptance of the idea of "setting the software free" proposed by Richard Stallman so that it is not locked away in heavily guarded vaults like valuables or jewels (closed source software) are by their owners or customers. In his opinion "source code should be shared like the air in a room" (Šoka 2011, 3). This decision by programmers subsequently led to the creation of internal cultural practices and rules within their community such as free acquisition and access to software and its collective and voluntary improvement. In addition to increasing the level of reliability of programs, these improvements bring participants an irreplaceable feeling of pleasure that accompanies the performance of difficult tasks without the unnecessary obstacles of constantly overcoming trivial pitfalls such as the poor logistics of storing solved tasks and the inconvenient and lengthy searching and sorting of data.

The motivations that accompany the members of a culture built upon the foundations of open sources software have been summarized by the Slovak researcher Milan Šoka:

> The more efficiently you can work on things – and I do not just mean software development, but in general – the more you can immerse yourself in them. The more you can immerse yourself in your work, the more you can develop a relationship with it. The more of a relationship and

love you have for your work, the more enjoyable it becomes and the more you want to pursue it next time. In addition, you also get the motivation to talk about your successes, share your experiences with others, and inspire others to try working the way you do. In my opinion, this process is self-fulfilling and self-expanding. [...] Perhaps it is because of this and versioning tools that Linux has become so successful, as has open sources software in general.
(Šoka 2011, 4)

The culture of giving, sharing, exchanging, and collaborating created the precondition for an atmosphere conducive to dialogue, the moderation of opinions, and the elimination of the aforementioned comprehensible dispute between programmers and literary scholars. From the perspective of interdisciplinary methodology, the inability to understand monodisciplinary discourses could be mitigated and levelled through the use of the interdisciplinary concept of exchange (Miller 1982; Homans 1974). For example, this could be in the form of methodological exchange – that is, making specific programs (source codes) available to scholars in the humanities and the arts. Specific programs have sometimes been provided to literary and arts scholars and have generated contributions in the form of innovative literary research conducted upon the basis of information theory, analytical statistical methods, computer modelling, and quantitative methods (Miller 1982). This is known as the "digital humanities".

As a result of initiatives around free software resources, the drive for an extreme (and clearly marked) autonomy in the discourse on electronic literature is now retreating. A more sophisticated view – whereby general digital literacy itself is less important than its differentiation and more precise definition – is coming to the fore. The initiative of David M. Berry, who works directly in the programming environment (*The philosophy of Software: Code and Mediation in the Digital Age*, 2011), and the efforts of Mark C. Marino, who is an academic teacher of literature and an author of art projects, should be seen as part of this tendency. Marino was probably the first to outline a perspective for defining the subject and method of the (sub)discipline of "Critical Code Studies" (2006). Berry complemented this by defining the specific skills necessary for practitioners and researchers in this new field and underlining the necessity of its related subdisciplines such as software studies and new media studies.

The logic of ontology

"A logic of ontology: an orientation in interdisciplinary practices towards effecting ontological transformation in both the object(s) of research and relations between the subjects and objects of research." (Barry and Born 2013, 249)

Critical code studies

On the way towards an interdisciplinary integration of otherwise disparate results, the starting point for Marino's search for a reconciliation between engineering and literary science approaches, which can be understood as an interdisciplinary practice, was a dual experience: the hermeneutic experience of interpreting literary works and the use of source codes that programmers create with a particular programming language (e.g. Java). The latter contains fully fledged commands such as "Print" and "Go", and characters composed of letters, numbers, and symbols. It also contains various operations (e.g. naming, comments, loops, and recursion) that characterize a set of strategies of particular processes that may be occasionally reused in any particular program (such as a phrase, a collocation, or a syntactic rule in natural language). It is therefore an artificial language with specific expressive possibilities. It is essentially the formulation of instructions in an artificial hybrid language which has to be learned, just as is the case with writing and reading. Let us put aside for the moment the object and the goal of notation – that is, how the program will (or will not) solve an assigned task. There is an established term for a written program – source code (or simply "code"). It is from this term that the name for critical process description – how specific programs are practically used; how they travel between authors; what purpose they were written for; what purpose they ultimately serve; and how they can be improved, made more efficient, broken, or abused – is derived. In summary, knowledge of at least a few programming languages is required to critically describe a source code. As a program written in a language, code can be seen as a semantic and semiotic unit; it can be subjected to a process of reading, understanding, and interpretation, just like an ordinary written text. In order to be able to do this, however, knowledge of interpretation processes is essential. According to Umberto Eco, like with the reading of literary texts, a computer program can also be misinterpreted by the reader (Marino 2006). Based on the above, it is clear why, in Marino's opinion, source code can (and should) be read in a similar way to a literary work. The object of the reading process is not the content itself (i.e. the task solved by the program or the target it is supposed to achieve) but rather the structuring and context of the program itself (Cayley 2012).

Based on the initial reactions of programming scholars, who strongly disregarded practices that were widespread in the humanities disciplines, the shift towards a certain openness to interpretive methods in the workings of "code" is surprising. They found it beneficial to commit the subject of their discipline to an ontological transformation and to begin to view source codes – like any other texts

– through the prism of hermeneutic practices. As writings recorded in hybrid artificial languages, programs therefore need to be described, analyzed, and interpreted, as well as explained, especially for the sake of ensuring that they do not affect life in an uncontrolled and exclusively deterministic or causal manner. They must be interpreted as if they were not merely a close reading strategy but rather as if they were mainly about revealing contexts and "extra-literary" aspects, identifying in turn the author of specific programs and analyzing the orders for specific solutions, the history of programming languages, and malicious programs (viruses, Trojan horses, logic bombs, and hoaxes), deliberately introduced bugs, and the consequences that result from them. At an online conference on critical code studies, Marino used the example of the misuse of a sportswear advertising program with photos of Anna Kurnikova. He highlighted the need to broaden the subject matter of the academic subdiscipline and include the psychological, cultural, social, and subjective aspects of the effects of source code.

The issue of computer literacy has been more specifically addressed by Berry, who added new particular dimensions to original conceptions of digital literacy (Berry 2011) linked to education (digital *Bildung*) under the influence of Marino's own reflections. He frames them with the term "iteracy" (Berry 2011), which he uses to refer to abilities and skills parallel to the abilities and skills that language users need to have in order to understand a text – that is, the literacy and mathematical skills necessary to work with numbers in different contexts.

The term "iteracy" is a synthesis of the lexemes "literacy" and "iteration", and it focuses on the practical skills necessary for using programming code languages, and ultimately reading and writing the programs created in them (code). On his specialized blog, David Berry has summarized the areas that he believes the term "iteracy" covers:

> Computational Thinking: being able to devise and understand the way in which computational systems work to be able to read and write the code associated with them [...]; Algorithms: understanding the specifically algorithmic nature of computational work, e.g. recessions, iteration, discretisation, etc.; Reading and Writing Code: practices in reading/writing code require new skills to enable the reader/programmer to make sense of and develop code in terms of modularity, data, encapsulation, naming, commentary, loops, recursion, etc.; Learning programming languages: understanding one or more concrete programming languages to enable the student to develop a comparative dimension to hone skills of iteracy, e.g. procedural, functional, object-oriented, etc.; Aesthetics of Code: developing skills related to appreciating the aesthetic dimension of code, here I am thinking of "beautiful code" and "elegance" as key concepts [...]; Data and Models: understanding the significance and importance of data, information and knowledge and their relationships to models in computational thinking; *Critical Code Studies: critical

approaches to the study of computer source code [...]; *Software Studies: critical approaches to the study of software (as compiled source code), particularly large-scale systems such as operating systems, applications, and games. (Berry 2011)

Such trajectories of thinking about code clearly point towards a hermeneutic perspective where code shall be analyzed and interpreted as text and thus as a system of signs with its own rhetoric and as verbal communication whose meaning transcends the particular functional use of code. Ultimately, Marino argues, we can read and explicate code as we would explicate a literary work in a new discipline of inquiry – namely, that of code criticism.

Forms of source code expression do exist

Based on Berry's concept, it is clear that this is a technologizing of the word *sensu stricto*. This is a view at the level of "below the surface of the monitor" and below the visible layer of the word (*visibilia*) and the perceptible materiality of the electronic sign. This is a layer that belongs to the level of textons (within Aarseth's concept of "cybertext") yet which also reaches into human consciousness in a very specific and almost uncontrolled way. From an everyday perspective, we are talking about a space where the design of each task that is solved by software takes place – as Ted Nelson defined it – as an interplay and a conceptual adequacy of the set task and the creative way of solving it (i.e. innovative solutions).[77] These are solutions that emerge within a certain routine habit (and possibility) of creating a (new) innovative structure of mind as a concept of dealing with the task.

Walter Ong also presented the technologizing of the word within the context of innovation as a long-term process that led to certain manifestations (effects) and to new structures of thought. Through writing, humans came to know about things they had not seen and ideas they had not heard about, thus arriving at abstract thinking and reasoning. These things could then start the process of penetrating into and affecting life. According to Ong, technology as a digital medium may intervene dramatically in the technology of writing, printing, and electronic formats, while doing the same more subtly and indirectly, less noticeably and dramatically, and intricately in the human consciousness (Ong 1982). This penetrates into life in a parallel way through a causal process, while some aspects and stages happen implicitly. Jan van Dijk (2006) explores the social aspects of new media with a similar approach.

Researchers more focused on the technically determined aspects of programmable media, believe that these open sources software-based media speed

[77] In these contexts of understanding the interface as a way of knowing reality, also see Hoffman (2009).

up the whole process but do not simplify it (Šoka 2011).[78] It seems that putting into practice specific programming solutions can directly or indirectly intervene in the lives of individuals and society in a myriad of ways, including economic, political, and psychological ones (Cox and McLean 2013). (In this context, the example of the recent economic crisis associated with virtual stock trading is most often cited.) This is especially the case because current strategies for writing programs (object-oriented programming) are aimed at minimizing the distance between actual reality and the virtual structure. (This is similar to the previous example about the recording of data in biological and artificial systems, where the perception of the distinction between artificial and living systems is minimized.)

It is clear that programs and software platforms – as they have been recently conceived and used by users of the Internet and of various applications – act as active players in relation to life (or being-in-the-world). Humans and machines read them on an equal level, hence the emerging view that source codes can also be spoken (Cox and McLean 2013); it is through source codes that humans communicate with each other and make decisions about themselves and about life. It is therefore quite justified to use the analogy of the active impact of the spoken word (oral culture) on life. But is this primarily a human or a posthuman consciousness? This uncertainty plays an important role in the process of interpreting writing in coding languages – in programs – that is, in quasi-texts. It ensures that the interaction between man and machine is not interrupted, that the interface between them is still the mediator, that the preconditions for their mutual communication remain present, and that man and machine interact. (This process is greatly complicated by deep learning technology.)

From Berry's concept, it is obvious that in the case of source codes (programs) there is indeed a technologizing of the word *sensu stricto*; however, the word does not lose its performative power, and it is in this technological mode and in the variety of goals that society achieves diverse effects and impacts then appear (Cox and McLean 2013; Chandra 2014). According to Marino, source code is most appropriately represented by the notion that it is the text of culture. Critical code studies are establishing the contextual interpretations of the content of programs. On the contrary, source codes produced with the aim of literary experiments (code poetry and codework) are not a part of the subject matter postulated by critical code studies.

---

[78] For more information, see: http://www2.fiit.stuba.sk/~bielik/courses/msi-slov/kniha/2012/Resources/Essays/Essay_86.pdf. Accessed June 27, 2021.

Marino argues that literary works that were created from source code, or source codes that were created for the sake of literary experiments, must be reinterpreted in other dimensions. This fact has not deterred the editors of *New Scientist* (Firth 2014) from presenting to its readership those source codes that scholarly literature has placed within the system of electronic literature and for which the "code poetry" and "codework" has been established. These are the source codes used by programmers for literary experimentation. In terms of the technologizing of the spoken word – the effect of which is electronic literature – it is important to observe how promoters and creators seek to establish and define code poetry within the context of working in the culture of symbolic meanings. The question of whether the process of creating codework deepens the alienation of contemporary man becomes crucial. Furthermore, what models and structures does it create? And how does it support the autonomy and closedness of the system of electronic literature? To what extent does code poetry merely refer to the notion of the interface? Or is it a legitimate analogy of it?

The logic of innovation

Poetry from the spirit of algorithms

It is worthwhile noting that this is currently only a discussion and not a concise conclusion of a discourse on varyingly binding procedures, ideas, schemes, and precise definitions, as was the case with discourses in literary science (Nycz 2002). The starting point for that direction was a constantly increasing material base as well as varying and dynamically changing methods. In the case of code poetry, the activity is of a performative nature. Its sequence, dynamics, and focus are determined by the effects of phenomena derived from the behaviour of users of programming languages, source code, and software platforms (how actively, willingly, and innovatively they use them, and to what extent they polemically discuss them). The origins of the polemics range from online discussions posted on professional websites, professional blogs, and online magazine discussions (essays, commentaries, and relevant analyses) through to video presentations posted on Vimeo, as was the case with Marino's discussions.[79]

The origins of code poetry can be found in earlier projects by new media artists (Talon Memmott and Mez; Mary-Anne Breeze) who were not averse to presenting works on the screen as a visual layer of source code which, in the language of programmers, is called a "textual interface". This is attractive in terms of the layout of lines on a screen, because it is associated with the visual rhythm

[79] See: http://vimeo.com/9124819. Accessed July 16, 2021.

of the particular poetry on the printed page. More telling was Themerson's idea of seeing the visual materials of scientific experiments as images (Reichardt, Wadley 2020). Other experiments – involving the hybridization of fragments of old programs with natural language, or featuring certain programming symbols and signs – also caused controversy among critics. For some time, there has been a requirement that a poem produced by source code must also be programmable. It can be assumed that it is for this reason that the interest in literary hypertexts among students has been replaced by the popularity of codework, which has been presented with great popularity, especially at slam poetry events.

It is clear from this summary of the empirically observed evolution of code poetry that it is a phenomenon with only an approximately defined boundary. The interest of the genre's promoters and critics centres on two problematic areas: Where is the evidence that it is poetry? If it is not poetry, what is it?

These questions point to the need to list arguments against categorizing code poetry as poetry as such. One of the first arguments is the attempt to define code as a language, and more specifically as a poetic one. The second is mainly about following (or not following) conceptions of the essentialist definition of poetry. The loose discussion of views on source code in a linguistic context, outlined by Loss Pequeno Glazier in "Code as Language" (2006), has been met with more cautious thinking on the subject in John Cayley's "Time, Code, Language: New Media Poetics and Programmed Signification" (2012). For Glazier, source code, which provides an apparatus for inscribing thoughts and emotions, falls into the category of language as a tool of communication (analogous to the aforementioned views of the promoters of constructivism and media theory). Cayley, on the other hand, takes into account linguistic concerns; for him, source code is only a specific kind of language system ("code is a special type of language"), which, after all, only "resembles" language. The effects of using this kind of language in the creation of "codeworks" (Husárová 2012, 83) are recommended for the attention of computer scientists and for contemporary literary scholars (Cayley 2012, 312). In the above contexts, the ideas of Melissa Kagan (an enthusiastic promoter and organizer of slam poetry) that this is a new dimension in the evolution of linguistic means of expression need to be taken with a grain of salt (Kagan in Firth 2014).

The focus of experts on linguistic and extralinguistic means of expression shifts their discussion towards a comparative perspective. The recurring premise that "code is poetry" is explained on a website by the platform's correspondent as simply being the usage of language on the Internet. The website points out that this is not a real analogy but rather a metaphorical reference to basic programming strategies. A more challenging task was tackled by Matt Ward (2010), a student of English literature and a practical designer, in "The Poetics of Coding", which is a study which is recommended reading for students of code poetry courses. Using empirical material, Ward demonstrated that a more or less precise analogy can be found between programming styles and the choices available from source code strategies. At times, this is surprisingly very precise. Note, for example, the requirement to precisely follow the structure of an English sonnet and the analogous (i.e. precisely specified) hierarchy of possible source code procedures.[80]

By demonstrating a correspondence or analogy between the type of skills in the innovative and creative use of constraints and limitations identified in poetry and the typologically similar skills known from source code, Ward concedes that "perhaps code really is a form of poetry, and the coder a new kind of poet" (Ward 2010).

This opinion, however, does not exhibit the characteristics of a scientific premise; it is based only on empirical experience and reflects the idea of poetry in the consciousness of people today. On the other hand, it does stimulate an exchange of ideas. In polemics, it functions as part of a set of terminological memes associated with the specific activity of creating and reading source codes: "code is text", "code is language", and "code is poetry".

Perhaps more relevant in this respect is the position of Vikram Chandra, an author of several literary pieces and the critical work *Geek Sublime: The Beauty of Code, the Code of Beauty.* He strongly challenges the notion of there being an analogy between the creation of poetry and the writing of (quasi-texts) in source code (Chandra 2014). He presents "code poetry" as a skill in handling programming languages and as an inventive creation of source code as a functional and semiotic whole. In his view, despite the fact that works of code poetry (unlike particular source code) do not have an obligation to fulfil the goal that is set at the beginning of a program, they are not poetry. This is primarily because, in the process of "reading", the reader almost automatically recognizes the superficial content and perceives the written content as a primitive structure and as "texts" that do not create or generate symbolic meanings. This argumentation is based on the notion of *dhvani* (Chandra 2014, 199), which the Indian scholar Ánandavardhana clearly identified as the soul of literature. The inspiration for Ánandavardhana's theory came from grammarians' reflections on *sphót* (literally "being in bloom"),

---

[80] An example of code poetry: https://medium.com/s/art-of-code/on-code-and-poetry-a-conversation-5c7d0c19be00. Accessed August 9, 2021.

which reveals the meaning of a word consisting of individual syllables. The literary text expresses the unexpressed meaning, which is called *dhvani* and which is distinct from the constituents of the literary work, just as the syllables of a word are distinct from the overall meaning. *Dhvani* is therefore the implied meaning. Ánandavardhana compares this to the charm of a woman, which is distinct from her limbs even though it can be perceived in them. This meaning can only be perceived by those who are sufficiently "literarily sensitive" (Gáfrik 2012, 36). The category of the unspoken, untold, and implied meaning is like a lens where the core of Chandra's "dispute" with the proponents of code poetry (as poetry) is concentrated. Is this the core of the misunderstanding of two different environments?

In conclusion, it is important to say that renowned scholars in the field of electronic poetry (a term broader than code poetry) search for ideological and creative pendants for the world of electronic creation primarily in references to the initiators of the artistic and poetic avant-garde in order to emphasize its radical difference from the existing works of printed literature.

However, if a poem written in source code is poetry, and if we read and perceive such a text as a poetic utterance, then it is "poetry from the spirit of algorithms". On a theoretical level, this perspective seems to have been embraced by Glazier, a renowned new media scholar, who argues that source code is fundamental to the strategy of structuring digital poetry (Glazier in Morris 2006, 8). Programmers also emphasize the creative process: they stress the analogies between the poetic creative process and the creation of a program in source code. Moreover, they believe that this is a matching of strategies based on skills appropriate for the goal: dense and appropriately chosen procedural strategies and the elegant use of the expressive means of (programming) language. This is a kind of analogy to the categories of *proprium*, *aemulatio*, and *imitatio* which were popular in the Renaissance. This tendency towards the self-definition of empirical authors of source code is linked by critics to manifestations of maker culture (Silverman 2014) or creative "handymen". Recent generations of digitally literate contemporaries are responding to the economic crisis in relation to a wide variety of needs and interests with the potential for self-service by making various missing material objects on their own or printing them on 3D-printers. This also includes artistic activity, and most likely also poetic production. From the point of view of cultural history, an analogy with the category of "otium" is offered in the given context as a well-known activity of educated people in the Renaissance who, in addition to their usual duties (e.g. in feudal or ecclesiastical structures), indulged in intellectual and artistic activities which they pursued for pleasure and enjoyment. (For instance, this is how translations and adaptations of Latin works and political and historical essays were produced in Poland.) It is clear from the above facts that the continued technologizing of words in the context of experimentation with source codes – "code poetry" – acts as an expression of the subculture of the environment of programmers and figures of the new literacy ("iteracy"). This puts pressure on the "horizon" of the present, from (or against) which electronic literature scholars must situate themselves to formulate premises and provide arguments in their scholarly reasoning.

The visible interface

This concept refers to Steve Wozniak's idea of humans interacting with machines directly through computer screens via visible icons. Metaphorically, it follows a practice known from the history of art (from late antiquity through to the eighteenth century) of placing written references (words, phrases, names, dates, and data – i.e. *verba visibilia*) upon the surface of a statue, sculpture, or painting which thus hybridized the visual image with a "semantic enclave" (Wallis 1983, 191). Conversely, they made the written word visible in this way. According to the semiotician, the authorial intention of early twentieth-century artists was carried out with a similar intention and aimed to make perception more difficult, surprises the viewer, and provide some aesthetic embellishment and provocation. Pablo Picasso, Michael Chagall, Paul Klee, and Max Ernst used alphabetic and hieroglyphic script, quotations, numbers, fragments of iconic cultural texts, bits of newspapers, and other things to realize this goal (Wallis 1983). Inscriptions of legible and illegible signs, *papiers collés*, and the later collage effect (Leo Malet) deliberately emphasized the visuality and visibility of semantic signs. Writing as a conventionalized medium – as Bolter would say already "limpid" (*immediacy*) – could not draw attention to what was invisible to man. And the avant-garde author highlighted what was invisible and elusive (*visibilité: invisibilité*). The authors of electronic literature strive for a similar effect.

It should be remembered that many technological and IT solutions have preceded today's computer interface. The evolution of programmable machines went from the stage of huge cabinets, keyboards with strange markings, and large strips of punched paper towards shrinking machines and console outputs (the interface between the machine and the human). Over time, the punched strips and labels disappeared and the console began to resemble the keyboard of a typewriter. The desktop monitor appeared, and on it there were text commands and subsequently text and images as well. Information on the history of computers, the development of programming languages, and the improvement of hardware and software has been summarized and annotated by several researchers. They have infiltrated the popular consciousness either as vague ideas, images from science fiction films, or as descriptions familiar to us from cyberpunk literature. However, for scholars of media archaeology and electronic literature, this is not just a progressive simple continuation but rather a series of major generational changes most often discussed in terms of three stages that discordantly build upon one another. Graphical interface computers concluded the age of perforated strips and labels; inconvenient windows and computers trapped on pads have been replaced by ubiquitous computing ("ubicomp"), which, in addition to providing programmed convenience in replacing humans in repetitive and non-innovative activities, does sometimes seem restrictive and frustrating in that it causes a sense of alienation. An awkward opportunity for a new solution has thus appeared. According to the researcher Lori Emerson, the gap that would soften uncomfortable feelings of emptiness is being filled by electronic literature (Emerson 2014), which is taking on the role of antidote to the pitfalls brought about by the "ubicomp" – that is, the diffuse presence of computing devices (Slovak scholars in computer studies use the term "ubiquitous artificial intelligence", whereas others speak of the "electronic sphere" that surrounds the globe).

For everyday users, these devices act as a clear window or a "view from an open window" that is known from art history (Alberti). This is precisely because the visionaries and designers of comfortable graphical interfaces have strived to create personal computers as the simplest possible devices, for which the user does not need any trained skills. It is simply a matter of intuitive control and immediate communication between man and digital machine. In this sense, literature uses terms such as "natural user interfaces" and "organic user interfaces". For the user, a suitably chosen and programmed interface is essentially barely perceived and seems directly natural. However, for art this is unsuitable as it is aesthetically empty. It would be a repetition that is boring and uninteresting. The limited potentiality of the visual form of hypertexts – the stable structure of the interface and the computer as a black box with guarded software – was probably one of the reasons for the loss of popularity and the potential of hypertexts in the environments of the hopeful creators of electronic literature. Conversely, with the culture of free software, artistic activity has grown and manifested itself at the level of visible and perceptible interfaces such as in videos and tailored applications that authors use for artistic creation and which "only" refer to literature (Montfort 2014).

As can be seen in the work of Emerson, creators of electronic literature (often coming from developer backgrounds) have been designing various forms of interfaces since the early 1960s (Emerson 2014). In the role of users, they learned to operate them, and in parallel they deliberately made visible the interface that had already been transparent to them. In this way, in a practically designed interface and later on in a purposeful application, they deliberately aimed to create an aesthetically impressive effect – for example, there is the effect of deliberate mistakes already included in poetological dimensions (i.e. the poetics of the glitch). The authors of electronic literature have made a particular procedure (application or program) visible by using it in a different context and with a different and non-standard aim or intention. By changing the mode of use, a particular interface then drew attention to itself, just as the *verba visibilia* did by their potential possibility for semantic meaning referring to the inexpressible. In this context, Emerson speaks more of the poetic and aesthetic function of electronic signs (analogous to the strategy of the aesthetic function of language proposed by Roman Jakobson). Nonetheless, an analogy with the premise of the semiotics of art seems to be at least heuristically relevant. Its advantage, however, is that it takes into account both iconic and semantic aspects from the outset. It also builds on the insights of constructivists, who, in the context of media theory, state that it is appropriate to think of a system of a linguistic means of expression in terms of semiotic features and especially in terms of a means of expression that is typical of certain social practices within a collectively shared and specific field of knowledge. Specific linguistic means, after all, do not refer to a field of reality but rather to the very process of communication and to common sense (Schmidt 2006, 314), which is so essential to the reading of source code or of code poetry.

It is evident that electronic literature is presented by Emerson as an artistic (posthuman) response to technological solutions and a causal result that emerges, operates, and evolves in an environment of multimedia signs, and which, in the case of source code, also forms a closed system. This leads to a technologizing of literature that results in the use (and misuse) of all elements of a particular interface – be it an audiovisual or textual interface, or even one from animation. From this point of view, there is little point in distinguishing between visual and literary art. However, there is a purpose in talking about the hypermedia artefact that is produced as a result of disrupting, damaging, and introducing errors into a particular interface. Emerson justifiably groups this dominant authorial strategy into three areas against iPad apps and mobile devices, codework (creation in source code), and hypertext/Web. She says that all three areas can be faulted for a myriad of reasons: they certainly put up a lot of barriers for the recipient and are not user-friendly. This is art that makes inventive use of limitations (in this case, technological ones).

What antidote will cyberneticists develop? How will authors approach the poetics of the interface? One answer looms: being ubiquitous and accessible, the interface can participate in the process of tackling any artistic project and in the search for strategies of artistic expression and artistic intervention in life – be it social, scientific, political, or economic. In this way, according to Barry and Born, the idea of interdisciplinarity can be realized, and art-science (in the sense of the accumulation of the three logics of responsibility, innovation, and ontology) can create alternative connections between science, technology, and society.

Between the printed form of literary texts and the remediated texts in the electronic medium, the human factor acts as an interface which enables (mutual) penetration from and into the cybersphere and electronic literature (e.g. programs; source code; literary games; video; electronic art as informatics literature; building on prose and poetic forms and types; and the manipulation of language through programs, applications, remixes, and hybrid texts).

Shifting the focus from remediation to the aspect of consciousness change that accompanies the first and second cybernetic waves in reflecting on hypermedia artefacts opens up the possibility for integrating other phenomena and realities. This leads to viewing the process of the emergence of new interfaces of literature – such as the hypermedia artefact – as a phenomenon created upon the basis of the fluid and dynamic rules of a virtual environment within which a clearly defined and differentiated form of posthuman consciousness breaks through to the surface. As some contemporary philosophers have sought to demonstrate, it is also possible that, in addition to the correlation of biological and cybernetic components determining the human environment, there is an innovation of the perception of reality for which the computer interface is an essential analogy (Hoffman 2009). Will this be a significant phenomenon that is analogous to the change in the structure of the mind that Walter Ong described in relation to the technologizing of the spoken word?

Conclusion

The creators and authors of new interfaces of literature expect from their readers an awareness of new hybrid forms of artistic expression, often determined by a change in the intentionality or purpose of the use of specific literary forms and specific works, as well as some knowledge of the various dimensions, practices, and activities that accompany the contemporary processes of communication and common sense. As a methodological consequence of this initiated process, new subdisciplines such as visual studies, media studies, software studies, and critical code studies are emerging. These all compete strongly with literary studies and literary history. This is because electronic literature marginalizes the text and subordinates it to the audiovisual system. Furthermore, authors sharing new skills, such as iteracy, are attempting (and intending) to participate in the production of literature and art.

Code poetry cannot be looked at from the perspective of a single monodiscipline – be it applied computer science or literary studies. For the naming of creativity in a programming language, computer science has no terms of its own. It adopts them from established academic disciplines, such as literary science. In this sense, the boundaries between these disciplines are diffuse. A hermeneutic approach also permeates the boundary: books and texts produced in programming languages have to be understood, explained, and interpreted by users and researchers. This is due to the requirement that a poem should be programmatically functional and the fact that it is clear that it is not enough to apply code analysis on its own ("code studies", Barry) when interpreting code poetry. Rather, there is the need for a different approach (Marino); perhaps a hybrid hermeneutics should be applied to the text in machine language as well as to the text in natural poetic language.

The focus of understanding and interpretation is balanced between linguistic invention (analogous to natural language invention) and programming innovation. This is a situation analogous to the creations of that stream of bioart that uses biotechnology and living material (tissues, cells, and bacteria). The English scholar Vid Simoniti (2017) argues that the artist creating bioart either uses biotechnology on a trivial level, merely as a means of expression, or becomes so deeply immersed in learning about the discipline and exploring a particular problem that he or she ceases to create art projects and moves into the realm of the researcher. One would almost like to conclude this chapter by raising a new topic that has been marginalized within art-science – namely, the sustainability of art in the field of research and technology.

Acknowledgement
This chapter was written at the Institute of World Literature, Slovak Academy of Sciences, as a part of VEGA project 1/0563/18.

Bibliography

Aarseth, Espen. [1997] 2020. *Cybertext. Perspectives on ergodic literature*. Baltimore: Johns Hopkins University Press.

Apostel, Leo. 1972. *Interdisciplinarity: Problems of teaching and research in universities*. Paris: Centre for Educational Research and Innovation of the Organization for Economic Co-operation and Development.

Barney, Darin D. 2008. *Społeczeństwo sieci*. Trans. by Marcin Fronia. Warszawa: Sic Press.

Barry, Andrew, and Georgina Born. 2013. *Interdisciplinarity: Reconfigurations of the Social and Natural Sciences*. London: Routledge.

Barry, Andrew, Georgina Born, and Gisa Weszkalnys. 2008. "Logics of Interdisciplinarity." *Economy and Society* 37, 1: 20–49. DOI: https://doi.org/10.1080/03085140701760841.

Bernstein, Mark. 2002. *Storyspace 1*. Accessed November 10, 2020. http://www.markbernstein.org/papers/HT02.pdf.

Berry, David M. 2011. "Iteracy: Reading, Writing, and Running Code." *Stunlaw. A critical review of politics, arts and technology*. Accessed December 15, 2020. https://stunlaw.wordpress.com/2011/09/16/iteracy-reading-writing-and-running-code/.

Berry, David. M. 2011a. *The Philosophy of Software: Code and Mediation in the Digital Age*. London: Palgrave Macmillan.

Bolter, Jay David, and Richard Grusin. 1999. *Remediation: Understanding New Media*. Cambridge, MA: The MIT Press.

Brillenburg Wurth, Kiene. 2017. "Remediation." In *Literature: An Introduction to Theory and Analysis*, ed. by Mads Rosendahl Thomsen, Lasse Horne Kjældgaard, Lis Møller, Dan Ringgaard, Lilian Munk Rosing, and Peter Simonsen. London, New York: Bloomsbury.

Castells, Manuel. 2008. *Społeczeństwo sieci*. Trans. by Mirosława Marody. Warszawa: PWN.

Cayley, J. Time. 2012. "Code, Language: New Media Poetics and Programmes Signification." In *New media poetics. Contexts, technotexts, and Theories*, ed. by Adalaide Morris and Thomas Swiss. Cambridge, MA: The MIT Press.

Cox, Geoff, and Christopher A. McLean. 2013. *Speaking Code: Coding as Aesthetic and Political Expression*. Cambridge, MA: The MIT Press.

De Kerchove, Derek. 2001. *Inteligencja otwarta. Narodziny społeczeństwa sieciowego*. Trans. by Agnieszka Hildebrandt and Roman Glegoła. Warszawa: Mikom.

Emerson, Lori. 2014. *Reading Writing Interfaces. From the Digital to the Bookbound*. Minneapolis: University of Minnesota Press.

Firth, Niall. 2014. "Rhyme and reason: Writing poems in computer code." *New Scientist*. Accessed November 10, 2016. https://zephr.newscientist.com/article/2014148-rhyme-and-reason-writing-poems-in-computer-code/.

Gáfrik, Róbert. 2012. *Od významu k emóciám. Úvaha o prínose sanskritskej literárnej teórie do diskurzu západnej literárnej vedy*. Trnava: Typi Universitatis Tyrnaviensis.

Glazier, Loss P. 2006. "Code As Language." *Leonardo* 14, 5. Accessed November 25, 2016. https://www.leoalmanac.org/wp-content/uploads/2012/09/01Code-As-Language-by-Loss-PequenI%CC%80%C2%83o-Glazier-Vol-14-No-5-6-September-2006-Leonardo-Electronic-Almanac.pdf.

Groth, Camilla, Pirjo Kääriäinen, Margherita Pevere, and Kirsi Niinimäki. 2019. "When Art meets Science: Conditions for Experiential Knowledge Exchange in Interdisciplinary Research on New Materials." *Knowing Together: Experiential Knowledge and Collaboration*. Accessed June 15, 2021. https://www.researchgate.net/publication/336086885_When_Art_meets_Science_Conditions_for_experiential_knowledge_exchange_in_interdisciplinary_research_on_new_materials.

Groth, Camilla, Pirjo Kääriäinen, Margherita Pevere, and Kirsi Niinimäki. 2020. "Conditions for Experiential Knowledge Exchange in Collaborative Research Across the Sciences and Creative Practice." *CoDesign* 16, 4: 328–344. https://doi.org/10.1080/15710882.2020.1821713.

Hayles, Katherine N. 2002. *Writing Machines*. Cambridge, MA: The MIT Press.

Hejmej, Andrzej. 2013. "Intermedialność i literatura intermedialna." In *Komparatystyka. Studia literackie – studia kulturowe*, ed. by Andrzej Hejmej, 97–121. Kraków: Universitas.

Hejmej, Andrzej. 2014. "Literatura w społeczeństwie medialnym." *Teksty Drugie* 2: 239–251. Accessed November 10, 2020. http://rcin.org.pl/Content/59786/WA248_79451_P-I-2524_hejmej-literat_o.pdf.

Higgins, Dick. 2000. "Intermedia." In *Nowoczesność od czasu postmodernizmu oraz inne eseje*. Trans. by Maria and Tomasz Zielińscy. Gdańsk: Słowo/obraz/terytoria.

Hoffman, Donald D. 2009. "The Interface Theory of Perception: Natural Selection Drives True Perception to Swift Extinction." In *Object Categorization. Computer and Human Vision Perspectives*, ed. by Sven J. Dickinson, Aleš Leonardis, Bernt Schiele, and Michael J. Tarr, 148–166. Cambridge: Cambridge University Press.

Homans, George C. 1974. *Social Behavior. Its Elementary Forms*. New York: Harcourt Brace Jovanovich.

Husárová, Zuzana. 2012. "O materialite elektronickej literatúry." In *V sieti strednej Európy: nielen o elektronickej literature*, ed. by Bogumiła Suwara and Zuzana Husárová. Bratislava: SAP, Ústav svetovej literatúry SAV.

Chandra, Vikram. 2014. *Geek Sublime: The Beauty of Code, the Code of Beauty*. Minneapolis: Graywolf Press.

Jantsch, Erich. 1972. "Towards interdisciplinarity and transdisciplinarity in education and innovation." In *Problems of teaching and research in universities*, ed. by Leo Apostel, 97–121. Nice: CERI/OECD.

Klein, Julie T. 1990. *Interdisciplinarity: History, Theory, and Practice*. Detroit: Wayne State University Press.

Kluszczyński, Ryszard W. 2001. *Społeczeństwo informacyjne. Cyberkultura. Sztuka multimediów*. Kraków: Rabid.

Kluszczyński, Ryszard W. 2011. *Towards the Third Culture. The Co-existince of Art, Science and Technology*. Warszawa: Narodowe centrum kultury.

Landow, George. 1992. *Hypertext: The Convergence of Technology and Contemporary Critical Theory*. Baltimore: Johns Hopkins University Press.

Marino, Mark C. 2006. "Critical Code Studies." In *Electronic Book Review*. Accessed November 10, 2019. http://electronicbookreview.com/essay/critical-code-studies/.

Marino, Mark C. 2013. "Readingexquisite_code: Critical Code Studies of Literature." In *Comparative Textual Media: Transforming the Humanities in the Postprint Era*, ed. by Katherine Hayles and Jessica Pressman. Minneapolis: University of Minnesota Press.

Miller, Raymond C. 1982. "Varieties of interdisciplinary approaches in the social sciences." *Issues in Integrative Studies* 1: 1–37.

Miller, Raymond C. 2001. "Beyond boundaries in international studies: A review." *Association for Integrative Studies. Newsletter* 1: 6–7.

Miller, Raymond C. 2010. "Interdisciplinarity: Its Meaning and Consequences." In *Oxford Research Encyclopedia of International Studies*. Accessed August 14, 2021. https://doi.org/10.1093/acrefore/9780190846626.013.92.

Montfort, Nick. 2014. "Nowe maszyny powieściowe: Nanowatt i Zegar światowy." *Ha!art!* 46. Accessed August 14, 2021. https://issuu.com/korporacja_haart/docs/ha_46_fragmenty/5.

Nicolescu, Basarab. 1985. *Nous, la particule et le monde*. Paris: Le Mail.

Nowotny, Helga, Peter Scott, and Michael Gibbons. 2001. *Re-Thinking Science: Knowledge and the Public in the Age of Uncertainty*. Cambridge: Polity.

Nycz, Ryszard. 2002. "Literatura nowoczesna: cztery dyskursy (tezy)." *Teksty Drugie* 4, 78.

Ong, Walter. 1977. *Interfaces of the Word*. Ithaca: Cornell UP.

Pang, Alex Soojung-Kim. 1998. "Hypertext, the Next Generation: A Review and Research Agenda." *First Monday* 3, 11. Accessed July 14, 2021. https://doi.org/10.5210/fm.v3i11.628.

Rankov, Pavel. 2006. *Informačná spoločnosť. Perspektívy, paradoxy, problémy*. Levice: LCA.

Reichardt, Jasia, and Nick Wadley (eds.). 2020. *The Themerson archive catalogue. Vol. 3. Gaberbocchus*. Cambridge, MA: The MIT Press.

Ryan, Marie-Laure, and Jan-Noël Thon. 2014. *Storyworlds Across Media: Toward a Media-Conscious Narratology*. Lincoln: University of Nebraska Press.

Rybson, Piotr. 2000. *Książki i Strony. Polska książka awangardowa i artystyczna w XX wieku.* Kraków: Karakter.

Scheliga, Kaja. 2014. "Interview with David M. Berry at Re:publica." *Stunlaw. Philosophy and Critique for a Digital Age.* Accessed November 10, 2020. http://stunlaw.blogspot.sk/2014/10/interview-with-david-m-berry-at.html.

Schmidt, Sigfried. 2006. "Konstruktywizm jako teoria mediów." In *Konstruktywizm w badaniach literackich*, ed. by Erazm Kuźma, Andrzej Skrendo, and Jerzy Madejski. Kraków: Universitas.

Schmidt, Sigfried. 2010. *Literaturoznawstwo jako projekt interdyscyplinarny. Teksty Drugie No 4.* Trans. by Bogdan Balicki. Warszawa: IBL PAN.

Silverman, Jacob. 2014. "Is Computer Coding an Art? Coders are makers. But what exactly are they making?" *The New Republic.* Accessed August 10, 2021. https://newrepublic.com/article/119215/geek-sublime-vikram-chandra-review-coding-art.

Stock, Paul, and Rob J. F. Burton. 2011. "Defining Terms for Integrated (Multi-Inter-Trans-Disciplinary) Sustainability Research." *Sustainability* 3, 8: 1090–1113. DOI: https://doi.org/10.3390/su3081090.

Šoka, Milan. 2011. "Nástroje, ktoré tvarovali softvér komunity otvoreného zdrojového kódu." In *Manažment projektov softvérových a informačných systémov.* Accessed June 10, 2020. http://www2.fiit.stuba.sk/~bielik/courses/msi-slov/kniha/2012/Resources/Essays/Essay_86.pdf.

Toomey, Anne H., Nils Markusson, Emily Adams, and Beth Brockett. 2015. "Inter- and Transdisciplinary Research: A Critical Perspective." In *GSDR Brief.* Lancaster Environment Centre. Accessed June 14, 2021. https://sustainabledevelopment.un.org/content/documents/612558-Inter-%20and%20Trans-disciplinary%20Research%20-%20A%20Critical%20Perspective.pdf.

Tribe, Mark. 2009. *New Media Art.* London: Taschen.

Van Dijk, Jan. 2006. *The Network Society: Social Aspects of New Media.* London: SAGE.

Vesna, Victoria. 2011. "Science Labs as Artist Studios". In *Towards the Third Culture. The Co-existence of Art, Science and Technology*, ed. by Ryszard W. Kluszczyński. Warszawa: Narodowe centrum kultury.

Wallis, Mieczysław. 1983. *Znaki i sztuki. Pisma semiotyczne.* Warszawa: PWN.

Ward, Mark. 2010. "The poetics of Coding." *Smashing Magazine.* Accessed August 14, 2020. http://www.smashingmagazine.com/2010/05/05/the-poetics-of-coding.

Yetiv, Steve A., and Patrick James (eds.). 2017. *Advancing Interdisciplinary Approaches to International Relations.* Cham, Schwitzerland: Palgrave Macmillan.

Zielinski, Siegfried. 2013. [...*After the media*]: *News from the Slow-Fading Twentieth Century.* Trans. by Gloria Custance. Minneapolis: University of Minnesota Press.

List of Contributors

Beňová, Barbora. The Department of Medical Ethics and Humanities, Second Medical Faculty of Charles University in Prague, Czech Republic.

Čana, Tomáš. The Department of Philosophy and Applied Philosophy, Faculty of Arts, University of Ss. Cyril and Methodius in Trnava, Slovak Republic.

Doležal, Adam. The Department of Medical Ethics and Humanities, Second Medical Faculty of Charles University in Prague, Czech Republic.

Ivanová, Kateřina. The Department of Public Health, Faculty of Medicine and Dentistry of Palacký University in Olomouc, Czech Republic.

Lemrová, Adéla. The Department of Public Health, Faculty of Medicine and Dentistry of Palacký University in Olomouc, Czech Republic.

Odorčák, Juraj. The Department of Philosophy and Applied Philosophy, Faculty of Arts, University of Ss. Cyril and Methodius in Trnava, Slovak Republic.

Pisarski, Mariusz. The Department of Philosophy and Applied Philosophy, Faculty of Arts, University of Ss. Cyril and Methodius in Trnava, Slovak Republic.

Rozemberg, Andrej. The Department of Philosophy and Applied Philosophy, Faculty of Arts, University of Ss. Cyril and Methodius Trnava, Slovak Republic.

Suwara, Bogumiła. The Institute of World Literature, Slovak Academy of Sciences in Bratislava, Slovak Republic.

Škoda, Jaromír. The Department of Medical Ethics and Humanities, Second Medical Faculty of Charles University in Prague, Czech Republic.

Tomašovičová, Jana. The Department of Philosophy and Applied Philosophy, Faculty of Arts, University of Ss. Cyril and Methodius Trnava, Slovak Republic.

Zielina, Martin. The Department of Medical Ethics and Humanities, Second Medical Faculty of Charles University in Prague, Czech Republic.